Grade Four S

Aligned to Alberta Curriculum

Written by Tracy Bellaire, Krystal Lapierre, Diane Schlicting
and Andrew Gilchrist

The lessons and experiments in this book fall under 5 main topics that relate to the Alberta curriculum for Grade 4 Science – Topic A: Waste and Our World, Topic B: Wheels and Levers, Topic C: Building Devices and Vehicles that Move, Topic D: Light and Shadows and Topic E: Plant Growth and Changes. In each lesson you will find teacher notes designed to provide you guidance with the learning intentions, the success criteria, materials needed, a lesson outline, as well as provide some insight on what results to expect when the experiments are conducted. Suggestions for differentiation or accommodation are also included so that all students can be successful in the learning environment.

Education Station
www.educationstation.ca
1-877-TEACH 'EM
1-877-832-2436

Copyright © On The Mark Press 2016, updated with answer keys January 2021

Some material appearing in this book has been used in other published works, such as Physical Science Grade 1 (OTM2144), Earth and Space Science Grade 1 (OTM2152) and Life Sciences Grade 1 (OTM2160).

Published in Canada by:
On The Mark Press
Belleville, ON
www.onthemarkpress.com

Funded by the
Government
of Canada

OTM-2173   ISBN: 978-1-4877-0201-4 © On The Mark Press

AT A GLANCE

Skills: Science Inquiry

4.1 Investigate the nature of things, demonstrating purposeful action that leads to inferences supported by observations. Students will ask questions that lead to exploration and investigation and identify possible answers to questions from themselves and others.

4.2 Identify patterns and order in objects and events studied; and record observations, using pictures, words and charts, with guidance in the construction of charts; and make predictions and generalizations, based on observations. Students will carry out procedures, identify materials and how they are used and access with guidance information and ideas from sources.

4.3 Investigate a practical problem, and develop a possible solution. Students will identify applications of what has been learned; identify new questions that arise from investigation and development of solutions to problem; identify the purpose of an object to be constructed; use various strategies to complete tasks; build a structure with moving parts; communicate results of construction activities and evaluate the product in terms of accomplishing goals and opportunities for improvements.

Attitudes

4.4 Demonstrate positive attitudes for the study of science and to the applications of science in responsible ways. Students will show growth in acquiring and applying curiosity, inventiveness, perseverance, appreciation of the value of experience and observation, a willingness to work with others, a sense of responsibility for actions taken and respect for living things and environments with a commitment for their care.

Topic A: Waste and Our World

4.5 Recognize that human activity can lead to the production of wastes, and identify alternatives for the responsible use and disposal of materials. Students will identify plant and animal wastes and describe how they are recycled in nature, identify and classify wastes that result from human activity, describe alternative methods of disposal along with their advantages or disadvantages, distinguish between wastes readily biodegradable and those that are not, compare packaging in regards to advantages and disadvantages to waste and identify methods of waste disposal currently used within the local community. Students will identify wastes toxic to people and the environment, identify materials and processes that may decrease the amount of waste produced, identify ways in which materials can be reused or recycled, develop a flow chart for a consumer product indicating the source materials, final product, use and method of disposal. Students will identify actions that minimize the production of wastes as well as develop and implement a plan to reduce waste and monitor what happens over a period of time.

Topic B: Wheels and Levers

4.6 Demonstrate a practical understanding of wheels, gears and levers by constructing devices in which energy is transferred to produce motion. Students will explain and demonstrate how rollers can be used to move an object in a practical situation, compare where wheels and rollers are used, construct devices that use wheels and axles and demonstrate their use model vehicles, pulley systems or gear systems. Students will construct and explain the operation of a drive system that uses wheel-to-wheel contact, belt, chain or elastic or cogs and gears. Students will construct and explain the operation of a drive system that transfers motion from one shaft to another where the second in parallel or perpendicular to the first. Students will demonstrate ways to use a lever that applies a small force or movement to create a large force or movement, predict how changes in the size of a lever or the position of the fulcrum will affect the forces and movements involved and construct models of levers to explain how levers are involved in other devices.

 OTM-2173   ISBN: 978-1-4877-0201-4 © On The Mark Press

Topic C: Building Devices and Vehicles That Move

4.7 Construct a mechanical device for a designated purpose, using materials and design suggestions provided. Students will design and construct devices or vehicles that move or have moving parts, use simple forces to power or propel a device, use energy-storing and energy consuming components to cause motion.

4.8 Explore and evaluate variations to the design of a mechanical device, demonstrating that control is an important element in the design and construction of that device. Students will apply control mechanisms in the design and use of devices, compare designs to identify relative strengths and weaknesses, identify steps used to construct a device or vehicle, work cooperatively with other students and design and make several models of a device in order to evaluate each model constructively.

Topic D: Light and Shadows

4.9 Identify sources of light, describe the interaction of light with different materials, and infer the pathway of a light beam. Students will recognize that eyes can be damaged by bright lights and eyes should be properly protected, identify a wide range of sources of light including the sun, electric lights, flames and materials that glow and distinguish between objects that emit their own light from those that require external source of light to be seen. Students will demonstrate that light travels outward from a source and continues unless blocked by an opaque material, describe changes in the size and location of Sun shadows during different times of the day, distinguish and classify between material that is transparent, partly transparent and opaque and recognize that light can be reflected using reflecting surfaces. Students will recognize and show that light can be bent (refracted) through certain objects and recognize that light can be broken into colours and that different colours of light can be combined for a new colour. Students will describe and use a variety of optical devices.

Topic E: Plant Growth and Changes

4.10 Demonstrate knowledge and skills for the study, interpretation, propagation and enhancement of plant growth. Students will describe the importance of plants to humans and their environment as sources of food, shelter and maintaining the environment, identify and describe the general purpose of plant roots, stems, leaves and flowers and describe common plants on the basis of their characteristics and uses. Students will recognize that plant requirements for growth, including water, space, light, energy, and air vary from plant to plant and conditions such as temperature, humidity and nutrients may also be important to particular plants. Students will identify plants with special needs, recognize the variety of plant communities found in the local area, recognize that differences in plant communities relate to variations in the amount of light, water and other conditions and recognize that plants of the same kind have a common life cycle producing new plants similar but not identical to parent plants. Students will describe ways that various flowering plants can be propagated from seed, cuttings, bulbs and by runners; nurture a plant through one complete life cycle – from seed to seed; describe the care and growth of a plant nurtured identifying the life stages of the plant, conditions and requirements of the plant and reproductive structures of the plant. Students will describe different ways that seeds are distributed and recognize adaptations for different methods of distribution.

Taken from the Alberta Education Grade 4 Science Curriculum.

TABLE OF CONTENTS

AT A GLANCE 2

TEACHER ASSESSMENT RUBRIC 5

STUDENT SELF-ASSESSMENT RUBRIC 6

INTRODUCTION 7

LESSONS

Topic A: Waste and Our World

Waste 8

Managing Waste 16

Reduce, Reuse, Recycle! 25

Packaging and Waste 36

Answer Key 164

Topic B: Wheels and Levers

Pulleys and Wheels All Around Us 45

Pulley Systems 52

Gearing Up 57

Gears in Motion 62

Wheels in Motion 71

Levers 80

Answer Key 167

Topic C: Building Devices and Vehicles that Move

Powered Up Vehicles 88

The Catapult (Control! Control! You Must Learn Control!) 97

Answer Key 169

Topic D: Light and Shadows

What is Light? 103

Light Travels 109

The Colours of Light 114

Casting Shadows 120

Light and Protection 127

Answer Key 170

Topic E: Plant Growth and Changes

Plant Parts 131

What Do Plants Need? 145

Special Needs and Scattering Seeds 158

Answer Key 173

OTM-2173   ISBN: 978-1-4877-0201-4 © On The Mark Press

Teacher Assessment Rubric

Student's Name: _____   Date: _____

Success Criteria	Level 1	Level 2	Level 3	Level 4
Knowledge and Understanding Content				
Demonstrate an understanding of the concepts, ideas, terminology definitions, procedures and the safe use of equipment and materials	Demonstrates limited knowledge and understanding of the content	Demonstrates some knowledge and understanding of the content	Demonstrates considerable knowledge and understanding of the content	Demonstrates thorough knowledge and understanding of the content
Thinking Skills and Investigation Process				
Develop hypothesis, formulate questions, select strategies, plan an investigation	Uses planning and critical thinking skills with limited effectiveness	Uses planning and critical thinking skills with some effectiveness	Uses planning and critical thinking skills with considerable effectiveness	Uses planning and critical thinking skills with a high degree of effectiveness
Gather and record data, and make observations, using safety equipment	Uses investigative processing skills with limited effectiveness	Uses investigative processing skills with some effectiveness	Uses investigative processing skills with considerable effectiveness	Uses investigative processing skills with a high degree of effectiveness
Communication				
Organize and communicate ideas and information in oral, visual, and/or written forms	Organizes and communicates ideas and information with limited effectiveness	Organizes and communicates ideas and information with some effectiveness	Organizes and communicates ideas and information with considerable effectiveness	Organizes and communicates ideas and information with a high degree of effectiveness
Use science and technology vocabulary in the communication of ideas and information	Uses vocabulary and terminology with limited effectiveness	Uses vocabulary and terminology with some effectiveness	Uses vocabulary and terminology with considerable effectiveness	Uses vocabulary and terminology with a high degree of effectiveness
Application of Knowledge and Skills to Society and Environment				
Apply knowledge and skills to make connections between science and technology to society and the environment	Makes connections with limited effectiveness	Makes connections with some effectiveness	Makes connections with considerable effectiveness	Makes connections with a high degree of effectiveness
Propose action plans to address problems relating to science and technology, society, and environment	Proposes action plans with limited effectiveness	Proposes action plans with some effectiveness	Proposes action plans with considerable effectiveness	Proposes action plans with a high degree of effectiveness

Student Self-Assessment Rubric

Name: _____   Date: _____

Put a **checkmark** in the box that **best** describes you.

Expectations	Need to do Better	Sometimes	Almost Always	Always
I am a good listener.				
I followed the directions.				
I stayed on task and finished on time.				
I remembered safety.				
My writing is neat.				
My pictures are neat and coloured.				
I reported the results of my experiment.				
I discussed the results of my experiment.				
I know what I am good at.				
I know what I need to work on.				

I liked _____

_____

I learned _____

_____

I want to learn more about _____

_____

OTM-2173   ISBN: 978-1-4877-0201-4 © On The Mark Press

INTRODUCTION

The lessons and experiments in this book fall under 5 main topics that relate to the Alberta curriculum for Grade 4 Science – **Topic A:** Waste and Our World, **Topic B:** Wheels and Levers, **Topic C:** Building Devices and Vehicles that Move, **Topic D:** Light and Shadows, and **Topic E:** Plant Growth and Changes. In each lesson you will find teacher notes designed to provide you guidance with the learning intentions, the success criteria, materials needed, a lesson outline, as well as provide some insight on what results to expect when the experiments are conducted. Suggestions for differentiation or accommodation are also included so that all students can be successful in the learning environment.

Throughout the experiments, the scientific method is used. The scientific method is an investigative process which follows five steps to guide students to discover if evidence supports a hypothesis.

1. Consider a question to investigate. For each experiment, a question is provided for students to consider. For example, "Is water always a liquid?"

2. Predict what you think will happen. A hypothesis is an educated guess about the answer to the question being investigated. For example, "I believe that it can change from a liquid into a solid or a gas." A group discussion is ideal at this point.

3. Create a plan or procedure to investigate the hypothesis. The plan will include a list of materials and a list of steps to follow. It forms the "experiment".

4. Record all the observations of the investigation. Results may be recorded in written, table or picture form.

5. Draw a conclusion. Do the results support the hypothesis? Encourage students to share their conclusions with their classmates, or in a large group discussion format.

ASSESSMENT & EVALUATION:

Students can complete the Student Self-Assessment Rubric in order to determine their own strengths and areas for improvement. Assessment can be determined by observation of student participation in the investigation process. The classroom teacher can refer to the Teacher Assessment Rubric and complete it for each student to determine if the success criteria outlined in the lesson plan has been achieved. Determining an overall level of success for evaluation purposes can be done by viewing each student's rubric to see what level of achievement predominantly appears throughout the rubric.

MEETING YOUR STUDENTS' NEEDS:

Depending on the needs of the students in your class, the teacher may want to scan any Teacher Notes into a digital format. By doing this, no matter the reading abilities of your students, they will be able to access the information of the text. When appropriate, teachers are also encouraged to allow students to collaborate on as many activities possible. This allows all students to be successful without modifying the text significantly.

TOPIC A: WASTE AND OUR WORLD: WASTE

LEARNING INTENTIONS:

Students will identify waste that is a result of human activity versus plant and animal waste, and the ways waste is recycled and managed in nature. Students will classify wastes that result from human activity.

SUCCESS CRITERIA:

- define the meaning of waste and provide examples of it on our planet
- determine which waste is due to human activity vs. plant and animal waste
- describe how waste is managed in the natural world
- create an earthworm farm
- record observations about earthworm activity
- make a conclusion about the earthworm's role in waste management

MATERIALS NEEDED:

- ask each student to bring in a large wide-mouthed glass jar with a lid
- a copy of *What is Waste?* Worksheet 1 for each student
- a copy of *Human Activity vs Nature* Worksheets 2 and 3 for each student
- a copy of *A Tree in a Forest* Worksheet 4 for each student
- a copy of *The Earthworms Clean Up!* Worksheets 5 and 6 for each student
- dictionaries or access to the internet
- read aloud about plant and animal waste management (see suggestion in #4 of procedure section)
- soil such as sand and loam or topsoil (enough to fill large jars for each student)
- a hammer and a nail, masking tape, a jug of water, a few small cups
- earthworms (2 or 3 per student)
- vegetable or fruit scraps

PROCEDURE:

This lesson can be done as one long lesson, or divided into shorter lessons.

1. Explain to students that they will learn about waste. Give them Worksheet 1 and a dictionary, or allow them access to the internet. They will research the meaning of waste, then engage in a Think, Pair, Share activity with a partner to brainstorm some examples of waste they have seen in our world. A follow-up option is to come back as a large group and record their ideas on chart paper that could be used for a later activity.

2. Give students Worksheet 2 to sort their ideas of waste into two categories, these being waste caused by human activity and waste caused by plants and animals. Give students a clipboard to put Worksheet 2 on, and a pencil. Take them on a walk through the neighbourhood to look for signs of waste. They will add these to Worksheet 2.

3. Give students Worksheet 3 to complete. Once again they will engage in a Think, Pair, Share activity with a partner to brainstorm some examples of how the natural world manages its own waste. A follow up option is to come back as a large group and record their ideas on chart paper that could be used for a later activity.

4. Read *A Tree in a Forest* (Author: Jan Thornhill) to the students. Ask students to listen for examples of how nature manages its own waste. Record students' ideas on chart paper as you read through the story. Give students Worksheet 4 to complete.

5. Students will have an opportunity to create their own earthworm farms. Give them Worksheets 5 and 6, and the materials to create the farms. Read through the question, materials needed, and what to do sections on Worksheet 5 with the students to ensure their understanding of the task. Students will make and record observations of the earthworm farms as they are created and again 24 hours later. They will make a conclusion about the purpose earthworms have in nature's waste management system.

OTM-2173   ISBN: 978-1-4877-0201-4 © On The Mark Press

*As an activity to enhance the learning about the necessity of decomposition in the natural world, show students The Magic School Bus episode called "Meets the Rot Squad". Episodes can be accessed at www.youtube.com.

DIFFERENTIATION

Slower learners may benefit by partnering up with a faster learner to complete the Think-Pair-Share activities. Also, this type of pairing could be of benefit while these learners complete Worksheet 2.

For enrichment, faster learners could choose one of their illustrations on Worksheet 4 and re-create it on a larger paper. They can paint and display it on a bulletin board.

Name: _____

What is Waste?

What exactly is **waste**? Use a dictionary to find the definition of this word, or use the internet to research its meaning. Record your answer below.

Waste is _____

_____

_____

_____

_____

_____

Think, Pair, Share

With a partner, do some thinking and sharing of ideas about waste you have seen or heard about happening on our planet.

In the box below, record your ideas of waste.

OTM-2173   ISBN: 978-1-4877-0201-4 © On The Mark Press

Name: _____

Human Activity vs Nature

Look at the ideas of waste that you recorded on Worksheet 1. Which ones are caused by human activity? Which ones are caused by plants and animals in the natural world?

Sort It Out!

Using point form, sort your ideas in the chart below.

Waste Caused By Human Activity	Waste Caused By Plants and Animals

Name: _____

What To Do:

1. Take a walk in your neighbourhood. Look for signs of waste. Add them to your sorting chart on Worksheet 2.

2. Look back at your sorting chart. What produces more waste, human activity or plant and animals? Justify your answer.

Think, Pair, Share

With a partner do some thinking and sharing of ideas of how the plant and animal world manages its waste. Record your ideas in the box below.

OTM-2173   ISBN: 978-1-4877-0201-4 © On The Mark Press

Name: _____

A Tree in a Forest

Illustrate and describe some ways that nature managed its own waste in the story *A Tree in a Forest*.

Name: _____

The Earthworms Clean Up!

When an earthworm eats and then digests food such as grass, leaves, or other vegetation, its castings add nutrients back to the soil that plants need in order to grow. Some people make worm farms in order to make rich soil to add to their gardens at home. Let's give this a try!

You'll Need:

- a large jar with a lid
- masking tape
- 2 or 3 earthworms
- a hammer and a nail
- soil (a mix of loam and sand)
- vegetable or fruit scraps
- a small cup of water
- 2 large sheets of black construction paper

What To Do:

1. Fill the jar until it is three quarters full of loam soil. Add a thin layer of sand to the top of the soil.

2. Throw some vegetable or fruit scraps on top of the sand.

3. Moisten the soil with a little bit of water. Then add the earthworms.

4. Using the hammer and a nail, your teacher will make some holes in the lid of the jar. Put the lid on the jar to close it.

5. On Worksheet 6, draw your observations of the jar and its contents.

6. Cover the sides of the jar with black construction paper so that no light can get in. Earthworms like it dark!

7. Leave the jar for a day.

8. The next day, remove the paper. Observe the changes in the jar. Record them on Worksheet 6.

OTM-2173   ISBN: 978-1-4877-0201-4 © On The Mark Press

Name: _____

Let's Observe

Draw your observations of the jar and its contents.

This is what it looked like in the jar before it was covered up:	This is what it looked like in the jar after it was left for one day:

Explain the changes in the jar. What did the earthworms do?

_____

_____

_____

_____

_____

_____

_____

_____

MANAGING WASTE

LEARNING INTENTION:

Students will learn about different types of waste, describe alternative methods of disposal, identifying advantages and disadvantages of each. Students will distinguish between wastes that are biodegradable and those that are not. Students will identify kinds of waste that may be toxic to people and to the environment. Students will identify methods of waste disposal currently used within the local community.

SUCCESS CRITERIA:

- describe three main categories of waste
- distinguish between biodegradable waste and hazardous waste
- research and report on a method of waste disposal
- investigate how waste is managed in the local community

MATERIALS NEEDED:

- a copy of *Types of Waste* Worksheet 1 for each student
- a copy of *Biodegradable, Recyclable or Hazardous?* Worksheets 2, 3, and 4 for each student
- a copy of *Waste Disposal* Worksheet 5 for each student
- a copy of *Waste Management in Your Community* Worksheets 6 and 7 for each student
- pencil crayons, pencils, markers, chart paper
- clipboards (one per student)
- access to the Internet

PROCEDURE:

***This lesson can be done as one long lesson, or can be divided into shorter lessons.**

1. Using Worksheets 1, 2, and 3, do a shared reading activity with the students. This will allow for reading practise and learning how to break down word parts in order to read the larger words in the text. Along with the content, discussion of certain vocabulary words would be of benefit for students to fully understand the passage.

   Vocabulary words: landfills, biodegradable, recyclable, filtration, byproduct, inert, substance, sewage, atmosphere, hazardous, composting, infectious

2. Give each student a green, blue and an orange pencil crayon, and Worksheet 4. They will categorize waste as either biodegradable, recyclable or hazardous.

3. Lead the class in a discussion on how to manage waste. Create a visual display or mind web with the common methods of waste disposal – landfills, ocean dumping, incineration (burning), recycling and composting. Highlight examples to the students advantages, disadvantages, limits and consequences of each method. Give students Worksheet 5. They will access the internet to research a method of garbage disposal. Once students have described the method and determined the advantages and disadvantages of the method, they can share their findings with another peer or in a small group. This will allow for students to learn about other methods of garbage disposal.

   Note: for two quick reference websites on methods of waste disposal, go to **https://goo.gl/di3JWJ** or **https://goo.gl/tzWHdi**

4. Arrange a class outing to the local waste management site. Give each student a clipboard, pencil, and Worksheets 6 and 7. They will interview the site worker about waste management practises at the site. Answers to the questions can be recorded in point form or by using pictures for

OTM-2173   ISBN: 978-1-4877-0201-4 © On The Mark Press

the students who may have difficulty with written output. *If an outing is not an available option, invite a worker from the local waste management site to come and speak to the students.

DIFFERENTIATION:

Slower learners may benefit by partnering up with a faster learner to research a method of garbage disposal on Worksheet 5. An additional accommodation is to have these learners only complete Worksheet 7 on the outing to the waste management site, omitting Worksheet 6.

For enrichment, faster learners could write a report on local waste management practises using the information they gained from the interview with the site worker.

OTM-2173   ISBN: 978-1-4877-0201-4 © On The Mark Press

Name: _____

Types of Waste

Waste can be divided into three main categories: **solids**, **liquids**, and **gases**. Waste must be treated differently depending on the state it is in.

Gaseous Waste

Gaseous waste is the most difficult type of waste to manage. Solid and liquid waste can make gaseous waste. When solid waste rots, harmful gases are produced into the air. When liquid waste evaporates it goes into the atmosphere as gaseous waste. Once gaseous waste is in the atmosphere, there is very little that can be done to control it.

Liquid Waste

Liquid waste is more difficult to see and to manage. It is what is collected in sewers and drainage pipes, and what is sent down drains and toilets. Water is a very important substance, so it is very important that we are able to remove liquid waste from it. Filtration systems are used to do this at water treatment facilities.

Solid Waste

Solid waste is the most visible type of waste. It is what we throw out every day, what we see as litter in the streets and in water, and what we see in landfills. Solid waste is the hardest to get rid of because it takes the longest to break down and because there is so much of it in our world.

OTM-2173   ISBN: 978-1-4877-0201-4 © On The Mark Press

Name: _____

Biodegradable, Recyclable or Hazardous?

Biodegradable Waste

Biodegradable waste is any type of waste that can be degraded biologically, which simply means, broken down.

Most biodegradable waste is made up of plant and animal products that can be broken down by bacteria in soil. Composting is an example.

Waste produced by the human body is biodegradable. This type of waste is the byproduct of digestion, and it is usually collected as sewage.

Inert waste is waste that cannot be broken down, but is not harmful to the environment. Some examples are sand, concrete, and chalk.

Recyclable Materials

Recyclable materials are any materials that can be used again, usually after undergoing a physical or chemical change. Some examples are **plastics**, **metals**, **wood**, **paper**, and **glass**.

Name: _____

Biodegradable, Recyclable or Hazardous?

Hazardous Waste

Hazardous waste is any type of waste that can be harmful to living things and the environment.

Corrosive waste is a chemical that damages or destroys other living or non-living materials on contact. Batteries contain corrosive waste.

Medical waste is a type of hazardous waste. It cannot be treated as normal waste because it is usually bio-hazardous or infectious.

Radioactive waste is any waste that has radioactive chemical elements. For example, nuclear radioactive waste is extremely dangerous to living things and to the environment if not managed correctly. Toxic waste is a chemical waste that is poisonous to living organisms.

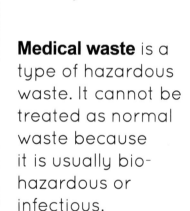

OTM-2173   ISBN: 978-1-4877-0201-4 © On The Mark Press

Name: _____

Categorize It!

Determine if the types of waste below are biodegradable, recyclable or hazardous.

Draw a green box around the biodegradable types of waste. Draw an orange box around the hazardous types of waste. Draw a blue box around the recyclable types of waste. Some pictures may receive more than one box around them.

Name: _____

Waste Disposal

Using the internet, research a method used to dispose of garbage.
Investigate and report on:

- the method of garbage disposal
- the advantages of this method
- the disadvantages of this method

Let's Investigate!

What method of garbage disposal do you want to investigate?

_____

Describe the **method** of garbage disposal:

_____

_____

_____

_____

Describe the **advantages** of this method:

_____

_____

_____

_____

Describe the **disadvantages** of this method:

_____

_____

_____

_____

OTM-2173   ISBN: 978-1-4877-0201-4 © On The Mark Press

Name: _____

Waste Management in Your Community

Visit your local waste management site and interview a municipal worker about how garbage is collected, sorted, and disposed for your community.

Let's Inquire!

1. How is garbage **collected**? How often is it collected?

_____
_____
_____
_____
_____
_____
_____
_____
_____

2. Is the waste **sorted** at the waste management site? How?

_____
_____
_____
_____
_____
_____
_____
_____

OTM-2173   ISBN: 978-1-4877-0201-4 © On The Mark Press

Name: _____

3. How exactly is the waste **managed** and **disposed** of?

_____

_____

_____

_____

_____

_____

_____

_____

4. What are the **advantages** of this method of waste disposal?

_____

_____

_____

_____

_____

_____

_____

_____

5. What are the **disadvantages** of this method of waste disposal?

_____

_____

_____

_____

_____

_____

_____

OTM-2173   ISBN: 978-1-4877-0201-4 © On The Mark Press

REDUCE, REUSE, RECYCLE!

LEARNING INTENTION:

Students will learn about ways in which materials can be reused and recycled, including examples of things that students use regularly. Students will identify actions that individuals and groups can take to minimize the production of waste, to recycle and reuse wastes and to ensure safe handling and disposal of wastes. Students will develop and implement a plan to reduce waste and monitor what happens over a period of time.

SUCCESS CRITERIA:

- identify objects that can be recycled or reused in order to reduce waste
- carry out plans of action to minimize waste (litterless lunch, composting, battery drive)
- make predictions, record observations, make conclusions about waste and reduction
- make a connection to the environment by identifying the benefits of waste reduction

MATERIALS NEEDED:

- a copy of *Reducing and Reusing* Worksheet 1 for each student
- a copy of *Recycling* Worksheet 2 for each student
- a copy of *Let's Go Litterless!* Worksheets 3 and 4 for each student
- a copy of *Classroom Composting* Worksheets 5, 6, and 7 for each student
- a copy of *The Life of a Battery* Worksheets 8 and 9 for each student
- recyclable materials such as a pop can, aluminum foil, paper, a cardboard box, a milk carton, a plastic juice jug, a glass jar
- a large plastic bin with a lid, a garden shovel, a drill, a long stick for stirring

- vegetable and fruit scraps, egg shells, leaves, grass clippings, used coffee grinds or tea bags, or other organic material, garden soil, access to water
- pencils, pencil crayons, markers, white poster paper (optional)

PROCEDURE:

*This lesson can be done as one long lesson, or divided into shorter lessons.

1. Discuss with students the meaning of recycling, reusing, and reducing. Using recyclable materials, demonstrate that these are recyclable and should be sorted. Give them Worksheets 1 and 2 to complete.

2. Explain to students that they are going to plan and advertise a litterless lunch day at school, where all students are encouraged to bring a lunch with minimal packaging in order to reduce the amount of waste that is produced at school. Give students Worksheets 3 and 4. They will design a poster to advertise the event, make a prediction of the outcome, observe, and make a conclusion about waste.

3. Working as a large group, students will create and contribute to a classroom composter. Give students Worksheet 5. Read through the materials needed and what to do sections with them to ensure their understanding of the meaning of composting and the task. Gather your materials and start building! Give students Worksheets 6 and 7. They will illustrate the composting layers at the start of the project, and make observations of the decomposition of the organic matter over time on Worksheet 6. After 8 weeks, students will illustrate the compost, and respond to questions on Worksheet 7.

4. Review with students the uses of a flow chart. Give students Worksheet 8 and 9. Read worksheet 8 with the class. On Worksheet 9 They will develop a flow chart for the production, use and disposal of retail batteries.

DIFFERENTIATION:

Slower learners may benefit by pairing up with a peer in order to design one poster that advertises a litterless lunch event, and creating one flow chart in the production, use and disposal of batteries. Some students may benefit from using flow chart software as an accommodation.

For enrichment, faster learners could write an announcement about the litterless lunch day and the battery drive event that could be read out during the school's daily announcements. After worksheets 8 and 9, students can design a poster for a battery drive to recycle batteries rather than disposing of them in landfill sites.

OTM-2173   ISBN: 978-1-4877-0201-4 © On The Mark Press

Name: _____

Reducing and Reusing

What exactly is reducing and reusing? Reducing is simply creating less waste to begin with, and reusing is using something for some other purpose that would have been otherwise treated as waste and disposed of.

Think, Pair, Share

With a partner, do some thinking and sharing of ideas of when you reduce and reuse. In the chart below, make a list of things that you reduce and reuse regularly.

Reduce	Reuse

Name: _____

Recycling

Recycling is an environmentally friendly way to dispose of objects that are no longer useful or needed. This helps the environment because new objects can be made from recycled materials.

Let's sort these objects into the recycling bins by drawing a line from each item to the correct recycling bin.

OTM-2173   ISBN: 978-1-4877-0201-4 © On The Mark Press

Name: _____

Let's Go Litterless!

We can also reduce our waste in the classroom in order to help the environment. Plan and advertise to have a "**Litterless Lunch**" day at school.

Let's Plan

Design a poster to advertise the event. Be sure to include some ideas on what would make good items to have in a litterless lunch.

[blank box for poster design]

What is another way you could advertise your litterless lunch idea at your school?

_____

_____

_____

Name: _____

Let's Predict

Make a prediction about what may happen during the litterless lunch.

_____

_____

_____

Now, carry out your plan, and observe if waste in your classroom has been reduced!

Let's Observe!

Describe what happened during your litterless lunch day event.

_____

_____

_____

_____

_____

_____

_____

Let's Conclude

Was your prediction correct? Explain.

_____

_____

_____

_____

_____

_____

OTM-2173   ISBN: 978-1-4877-0201-4 © On The Mark Press

Name: _____

Classroom Composting

Composting is a way to reduce the amount of waste that is dumped into landfill sites. By breaking down things like food waste, leaves, grass clippings, and wood bits into humus, we can put humus back into gardens, and reuse useful nutrients to help to grow healthy new plants.

Let's give back to the earth by creating some compost!

You'll Need:

- a large plastic bin with a lid
- leaves, grass clippings
- vegetable and fruit scraps
- a long stick for stirring
- garden soil
- a garden shovel
- a drill
- access to water
- eggshells

What To Do:

1. Put some soil in the bottom of the plastic bin.

2. Add some vegetable or fruit scraps, egg shells, leaves, grass clippings, and used coffee grinds or tea bags on top of the soil.

3. Moisten the soil with a little bit of water.

4. Repeat steps 1, 2, and 3 to make layers. Illustrate it on Worksheet 6.

5. Using the drill, your teacher will make some holes in the lid of the bin.

6. Put the lid on the bin to close it, and place it outside.

7. Use the long stick to stir your compost pile each week. Add organic materials when available. Make and record your observations of the changes you notice on Worksheet 6.

8. After several weeks, you will have created some humus that you can add to a garden to grow new plants. Illustrate it on Worksheet 7.

Name: _____

Draw a diagram of the composter. Show the layers of ingredients you put in it.

Let's Observe!

As organic matter "cooks", explain the changes that you are noticing.

_____

_____

_____

_____

_____

_____

OTM-2173   ISBN: 978-1-4877-0201-4 © On The Mark Press

Worksheet 7

Name: ______________________

How will you use your humus?

____
____
____
____
____
____
____
____

Illustrate the humus you have created.

Make a list of reasons why people should compost.

____
____
____
____
____
____

Challenge Question:

How does composting happen in nature?

____
____
____
____

Name: _____

The Life of a Battery

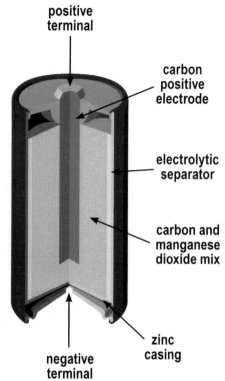

positive
terminal

carbon
positive
electrode

electrolytic
separator

carbon and
manganese
dioxide mix

zinc
casing

negative
terminal

Materials

Battery makers use all kinds of materials. **Copper**, **aluminum**, **acids** or **electrolytes** are used to make the inside of batteries. **Plastics** are often used to make the outside casing.

Production

To make batteries, you need energy! People design the battery and then use machines to assemble all the materials together.

Final Product

When you see batteries for sale in a store, they are wrapped in paper and plastic. That **packaging** eventually becomes recyclable waste.

Using the Batteries

Do you use something at home that needs batteries? **Electronic devices** like game consoles, flashlights and smoke alarms are just a few examples of things that use batteries.

End of Use

The electrolyte inside the battery can be **corrosive**. That means it can be toxic and cause burning to living and non-living things. When batteries reach the end of their use, the electrolyte inside them can sometimes leak out. If corrosive and toxic matter leaches out into the environment, plants and animals can be seriously affected.

This is why it is important to keep batteries out of our landfill sites and to collect and recycle them properly. Instead of dumping something corrosive into the environment, the recycling centres **neutralize** the electrolyte. This means they add chemicals to make it no longer corrosive or toxic.

OTM-2173   ISBN: 978-1-4877-0201-4 © On The Mark Press

Name: _____

It is your turn to make a flow chart of the life of a battery! Be sure your flow chart has a place for the five steps on Worksheet 8. Include the important vocabulary words and your own drawings as well.

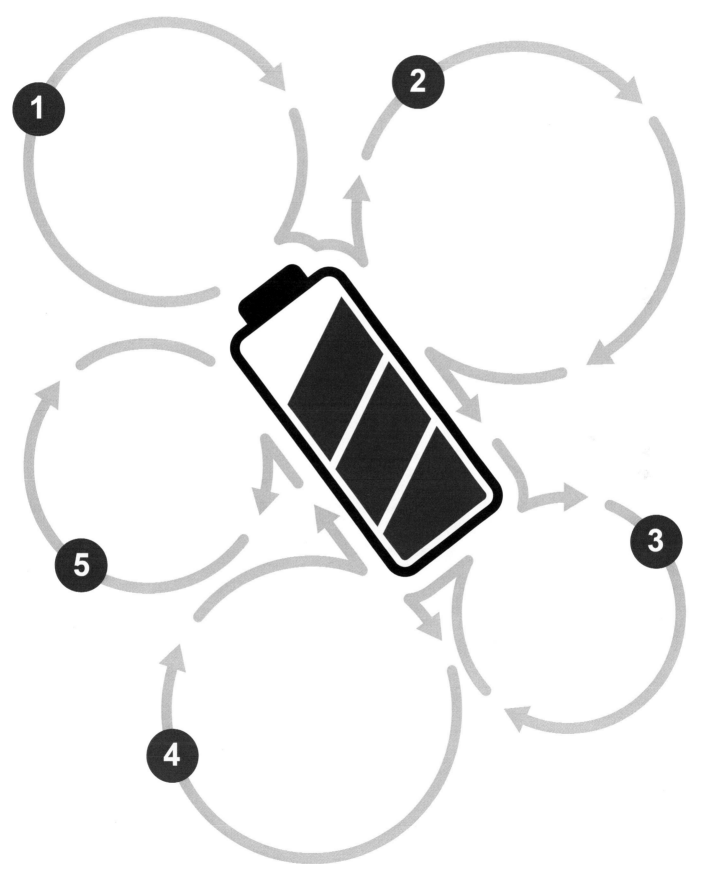

PACKAGING AND WASTE

LEARNING INTENTION:

Students will compare different kinds of packaging and infer the relative advantages and disadvantages of that packaging, considering a consumer perspective as well as an environmental perspective. Students will identify alternative materials and processes that may decrease the amount of waste produced.

SUCCESS CRITERIA:

- gather information and evaluate information based on criteria
- demonstrate ability to express advantages and disadvantages for consumers and environment while seeking to find solutions that maximize advantages and reduce disadvantages
- make observations and conclusions about the methods and benefits of waste reduction

MATERIALS NEEDED:

- a copy of *Pre-Consumer Waste* Worksheets 1 and 2 for each student
- a copy of *What Goes Into Packaging?* Worksheets 3, 4, and 5 for each student
- a copy of *Solving Problems – Environmental Packaging* Worksheets 6 and 7 for each student
- examples of packaging: cellophane wrapping, cardboard boxes, Styrofoam, bubble wrap, plastic trays, twist ties, foil
- packaged products – can include simple items such as bars of soap, take-out coffee cups, electronic devices, light bulbs, soft drink containers, etc.
- pencils, pencil crayons
- access to the internet or access to a local library (optional)

PROCEDURES:

*This lesson can be delivered as one longer lesson or divided into shorter lessons.

1. Lead a discussion on packaging materials with the students. Use the examples of packaging so that students can identify the material and think of examples of where they might see the material. Give students Worksheets 1 and 2. Read Worksheet 1 with the students. Check for understanding. Read through Worksheet 2 activities with the students. Do "Day 1" of the journal as a class activity so that you can either model the work or monitor their progress. Assign the rest of the journal as needed.

2. Tell students to bring from home an example of a product still in the packaging. Use the examples of packaging to help the students understand the different categories of materials – plastic, paper, cardboard, Styrofoam, metal and other. Tell the class they will be learning more about the impact of the packaging. (As an option, this activity and Worksheets 3, 4, and 5 can be led as a teacher demonstration or completed as groups where students share example products to fill out the chart.) With the class, use examples of packaged products and identify the materials used in the packaging. Does this have plastic in the packaging? Does this have cardboard? Can the packaging be reused? How about recycled? How could we reduce the amount of material used? Could we make the cardboard thinner or lighter? Could we make the box smaller or larger? Give students Worksheets 3, 4, and 5. Read through the instructions. If needed, use an example product to fill out the first line of the chart with students. Monitor student work as needed for the rest of the worksheet activities.

3. Give out Worksheets 6 and 7 to the students. Read Worksheet 6 with the students. Brainstorm some ideas for new products with the students, using examples as needed. Highlight some of the ideas

OTM-2173 ISBN: 978-1-4877-0201-4 © On The Mark Press

from the other worksheets to prime the students for thinking about packaging. Read Worksheet 7 with the students so they know what they will be responsible for after designing their product and packaging. Monitor their work.

DIFFERENTIATION:

Slower learners may benefit from a teacher-directed or group-directed journal on Worksheet 2. Pair slower learners with others in the class to give accommodations to slower learners. For Worksheets 6 and 7, prompt slower learners with prepared product and packaging ideas familiar to them – a soft drink in a can, an electronic device or a toy.

For enrichment, faster learners could research innovative packaging ideas such as biodegradable paper and cardboard made with embedded seeds or edible food holders for restaurants.

Name: _____

Pre-Consumer and Post-Consumer Waste

Pre-consumer waste is usually made up of raw materials or materials needed to create a product. Sometimes these materials aren't even in the product itself. For example, a company that makes newspapers would create the following pre-consumer waste:

- paper trimmings or misprinted papers
- ink containers, shipping boxes and wrapping material
- unsold newspapers

Post-consumer waste is anything purchased or used and then discarded.

Here are some examples of pre-consumer and post-consumer waste.

Pre-Consumer Waste	Post-Consumer Waste
Raw materials: wood, glass, metal, oil	**Packaging materials:** plastic wrap, bottles, boxes, styrofoam peanuts
Broken Products: bottles, bent cans	**Food waste:** peels, cores, bones
Byproducts: sawdust from wood, chemicals	**Used or outgrown items:** toys, clothing, magazines, calendars, bags, tissues
Transportation materials: plastic wrap, wooden pallets, string or ties	**Broken items:** electronics, light bulbs
Expired food products: spoiled vegetables or meat	**Cleaning solvents:** dish soap, laundry detergent, bleach

Consumer Perspective and Environmental Perspective

Packaging can effect what a consumer thinks about a product. The consumer may only want to buy a product that makes things easy for them. However, the consumer might avoid a product with a lot of heavy packaging or material that can't be reused or recycled.

When a product uses a lot of packaging or creates a lot of waste, it can be said to have an impact on the environment. It can cost a lot of money, time or effort to reuse, recycle or throw out the waste in a safe way. This is why many people look for ways to reduce packaging and waste.

OTM-2173 ISBN: 978-1-4877-0201-4 © On The Mark Press

Name: _____

Think about the things you throw out - in your lunch, in your desk at school, in your kitchen at home or when you are outside. Where did you put the waste? What do you think happens to it after that? Is it reused or recycled or put in a landfill?

Keep a journal of some of the things you throw out over 4 days. One of the things you throw out each day. Draw a picture and fill in the blanks for each day.

Day 1

I threw out _____

_____

I put it in _____

What do you think happened to it?

_____

_____

Day 2

I threw out _____

_____

I put it in _____

What do you think happened to it?

_____

_____

Day 3

I threw out _____

_____

I put it in _____

What do you think happened to it?

_____

_____

Day 4

I threw out _____

_____

I put it in _____

What do you think happened to it?

_____

_____

Pick one item from your journal. Can you think of a way you could reduce the packaging or waste you threw out? Explain.

_____

_____

OTM-2173   ISBN: 978-1-4877-0201-4 © On The Mark Press

Name: _____

What Goes Into Packaging?

Discarded packaging makes up a lot
of household garbage and recycling.
Almost every time you buy a new product
it comes packaged in something –
cellophane wrapping, cardboard boxes,
Styrofoam, bubble wrap, plastic
trays, twist ties or foil.

Let's examine the different kinds of
packaging.

What To Do:

1. Collect as many different packaged products as you can. Try to find
   packaging for food, electronics, cleaning supplies, school supplies and
   household products.

2. For each product, decide if you think the packaging can be reused
   somehow. Give an example of how and write your answer in the chart
   on Worksheet 4.

3. For each product, decide if you think the packaging can be reduced
   somehow. Give an example of how and write your answer in the chart
   on Worksheet 4.

4. For each product, decide if you think the packaging can be recycled
   somehow. Write your answer in the chart on Worksheet 4.

5. Separate the different materials in the packaging into piles – plastic,
   paper, cardboard, Styrofoam, metal and other. If some parts can't be
   separated by material, put them into the 'other' pile.

6. Answer the questions on Worksheet 5.

OTM-2173   ISBN: 978-1-4877-0201-4 © On The Mark Press

Name: _____

Product	Draw a picture of the item.	Draw a picture of the packaging.	Can the packaging be *reused?*	Can the packaging be *reduced?*	Can the packaging be *recycled?*

Name: _____

Think About It!

What product had the **least** amount of packaging?

_____

Do you think the packaging makes it easy or hard
for the consumer to buy the product? (*Circle one*)   **Easy**   **Hard**

Why? _____

_____

What product had the **most** amount of packaging?

_____

Do you think the packaging makes it easy or hard
for the consumer to buy the product? (*Circle one*)   **Easy**   **Hard**

Why? _____

_____

Look at your piles of packaging materials. Which one turned out to be
the **biggest**? (*Circle one*)

Plastic   Paper   Cardboard   Styrofoam   Metal   Other / Mixed

Which piles can be **reused** or **recycled**? *Circle* all that are *correct*.
Underline all that *cannot be recycled*.

Plastic   Paper   Cardboard   Styrofoam   Metal   Other / Mixed

---

Do you think packaging is necessary? Why or Why not?

_____

_____

_____

OTM-2173   ISBN: 978-1-4877-0201-4 © On The Mark Press

Name: _____

Designing Your Packaging

Imagine you are selling a new product. You need to design packaging for your new product. The packaging must protect the product so that it won't break or spill. But you also need to make it look interesting to your customer. And just as important, you need to make sure the packaging material has a low or no environmental impact.

My product is called _____

My packaging would use the following materials:

Plastic   Paper   Cardboard   Styrofoam   Metal   Other _____

Does your packaging include any special features? Explain.

_____

_____

_____

Name: _____

Think About It!

What does your packaging do to protect your product from breaking?

_____

_____

_____

_____

What does your packaging do to make your product look interesting to your customer?

_____

_____

_____

_____

What packaging material did you use so that it had a low environmental impact?

_____

_____

_____

_____

_____

OTM-2173   ISBN: 978-1-4877-0201-4 © On The Mark Press

TOPIC B: WHEELS AND LEVERS:
PULLEYS AND WHEELS ALL AROUND US

LEARNING INTENTION:

Students will learn that some wheel systems like pulleys are designed to lift a load and some pulley systems are designed to transmit power. Students will identify wheels in pulleys and show a practical understanding of the use of wheels in devices. Students will compare the wheel and the roller, and identify examples where each are used.

SUCCESS CRITERIA:

- identify everyday objects that use a wheel pulley system to lift a load, or use a pulley system to transmit power

- gather and record information using drawings and written descriptions

- classify pulley systems in objects as either designed to lift a load or as a belt drive that transmits power

- make observations, conclusions, and connections to people and places in the environment

MATERIALS NEEDED:

- a copy of **Pulleys All Around Us** Worksheets 1 and 2 for each student

- a copy of **Pulleys Everywhere** Worksheets 3 and 4 for each student

- a copy of **Comparing Wheels and Rollers** Worksheet 5 for each student

- clipboards (one for each student)

- a few (assorted) pulleys to use as examples for students

- a few (assorted wheels to use as examples for students – baking roller or wood doweling, toy wheels, bike wheels

- chart paper

- pencils, pencil crayons, markers

PROCEDURE:

***This lesson can be done as one long lesson, or divided into shorter lessons.**

1. Using the assorted pulleys as manipulatives, explain to students the purpose of the pulley. Have a brainstorming/discussion session as to what objects use a pulley system, or where students have seen a pulley system being used. Record responses on chart paper. Give students Worksheets 1 and 2 to complete, ensuring their understanding of the pulley's ability to lift a load or to act as a belt drive system that transmits power.

2. Explain to students that they are going to take a walk around the school to locate pulleys used in everyday objects. Instruct them to take note of the pulleys' ability to either lift a load or to act as a belt drive system that transmits power. Give each student a clipboard and Worksheet 3 to complete as they walk through the school.

3. Next, explain to students that they will take a walk around the neighbourhood to look for objects that use a pulley system to operate. Instruct them to take note of each pulley's ability to lift a load or to act as a belt drive system that transmits power. Give each student a clipboard and Worksheet 4 to complete as they go through the neighbourhood.

4. Come together as a large group and have a discussion about objects that use pulley systems to operate. (Prompt students to recall what they observed while on their walk through the school and neighbourhood.) Make a list of their responses on chart paper. How many had they thought of before their walks? How many did they come up with after? Were any the same?

5. Lead a discussion on wheels. Where do we use wheels? What do all wheels have in common? (a short cylinder, an axle, often attached to the machine). Introduce students to the term "roller." Use examples and compare the roller to the wheel. Are they both round in shape? Does the roller turn on an axle in its middle? Is it a short cylinder or a long cylinder? Give students Worksheet 5 to complete.

DIFFERENTIATION:

Slower learners may benefit from locating and drawing only two everyday objects in the school that use a pulley system, and only two objects in their neighbourhood that use a pulley system to operate.

For enrichment, faster learners could plan and carry out a field trip around their homes to look for objects that use a pulley system to operate. They could draw these objects and classify the pulley systems in them as either designed to lift a load or as a belt drive that transmits power.

OTM-2173   ISBN: 978-1-4877-0201-4 © On The Mark Press

Name: _____

Pulleys All Around Us

A pulley is a wheel with a groove in it. The groove holds a rope or a belt. Some pulley systems are designed to lift a load. A downward pull on one side of a pulley rope causes the opposite side to go up.

Circle the pulley in each picture, and explain how the pulley is being used.

_____

_____

_____

_____

_____

_____

_____

_____

_____

_____

_____

_____

Name: _____

Pulleys All Around Us

Some pulley systems are designed to transmit power. They are called "belt drives." A bicycle chain is a belt drive because it transmits power from the pedals to the back wheel.

Circle the belt drive in each picture, and explain how it is being used.

belt drive

_____

_____

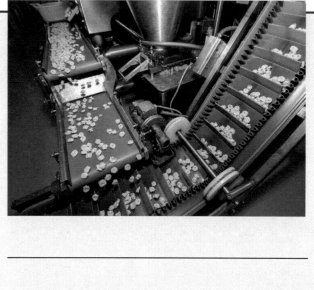

_____

_____

_____

_____

_____

_____

OTM-2173 ISBN: 978-1-4877-0201-4 © On The Mark Press

Name: _____

Pulleys Everywhere

Take a walk around your school to look for everyday objects that use a pulley system to operate. Draw and label four examples that you see.

Check (✓) one that applies:
- ❏ pulley lifts an object
- ❏ pulley has a belt drive that transmits power

Check (✓) one that applies:
- ❏ pulley lifts an object
- ❏ pulley has a belt drive that transmits power

Check (✓) one that applies:
- ❏ pulley lifts an object
- ❏ pulley has a belt drive that transmits power

Check (✓) one that applies:
- ❏ pulley lifts an object
- ❏ pulley has a belt drive that transmits power

Name: _____

Pulleys Everywhere

Take a walk around your neighbourhood to look for objects that use a pulley system to operate. Draw and label four examples that you see.

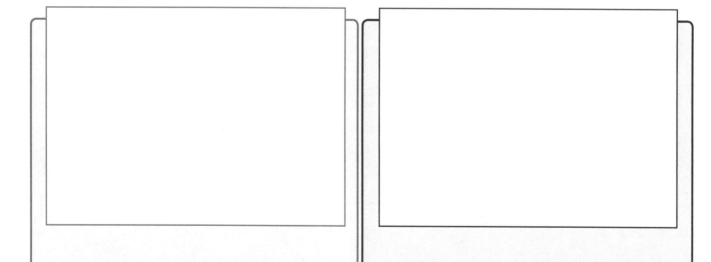

Check (✓) one that applies:

❑ pulley lifts an object

❑ pulley has a belt drive that transmits power

Check (✓) one that applies:

❑ pulley lifts an object

❑ pulley has a belt drive that transmits power

Check (✓) one that applies:

❑ pulley lifts an object

❑ pulley has a belt drive that transmits power

Check (✓) one that applies:

❑ pulley lifts an object

❑ pulley has a belt drive that transmits power

OTM-2173   ISBN: 978-1-4877-0201-4 © On The Mark Press

Name: _____

Comparing Wheels and Rollers

Circle the wheel or the roller in each picture. In each of the boxes, add a label with the word **wheel** or the word **roller**.

PULLEY SYSTEMS

LEARNING INTENTION:

Students will learn how wheel systems work in order to make lifting or moving objects easier. Students will construct devices that use wheels and axles to demonstrate and describe their use in pulley systems.

SUCCESS CRITERIA:

- identify and make a fixed pulley system and a block and tackle pulley system
- make a prediction, observations, and record effort ratings using a spring scale
- make conclusions about the advantages in the use of pulley systems
- make connections to people and the environment

MATERIALS NEEDED:

- pulleys with hooks (two per group of students)
- pails with handles (one per group of students)
- spring scales (one per group of students)
- rope or string
- weights 1 lb. to 2 lb. (or 500g and 1 kg)
- a copy of *The Fixed Pulley* Worksheets 1 and 2 for each student
- a copy of *Block and Tackle Pulley System* Worksheets 3 and 4 for each student
- pencils

PROCEDURE:

1. Explain to students that they are going to investigate how a pulley system works to lift a load. Divide students into small groups, and give them the materials they need to conduct both parts of the first experiment. Read through the question, materials needed, and what to do sections on Worksheet 1. Students will make a prediction about the outcome of the experiment, make observations through drawings, rate the effort required to lift the weight with and without the use of a pulley, then make a conclusion on Worksheet 2.

(The recorded numbers from the spring scale will be similar, but students should conclude that it is easier pulling down to lift an object, while using a pulley, than it is to lift upward.)

2. Introduce the block and tackle pulley system to the students by reading through the description on Worksheet 3. In their small groups, students will investigate the advantages of this pulley system. Give them Worksheet 3 and 4, and the materials they need to conduct both parts of the experiment. Read through the question, materials needed, and what to do sections on worksheet 3. Students will make a prediction about the outcome of the experiment, make observations through drawings, rate the effort required to lift the weight with the use of a simple fixed pulley and with a block and tackle pulley system, then make a conclusion on Worksheet 4.

DIFFERENTIATION:

Slower learners may benefit by using a pre-made block and tackle pulley system in order to conduct the experiment on Worksheet 3 and 4. This will eliminate a time consuming step, and allow them more time to focus on the task of comparing the effort needed to lift the weight.

For enrichment, faster learners could discuss in a small group, where they have seen the block and tackle pulley system used and how it was helpful in reducing the work effort needed.

OTM-2173   ISBN: 978-1-4877-0201-4 © On The Mark Press

Name: _____

The Fixed Pulley: Wheel and Fixed Axle

Question: Does a pulley make lifting easier?

Materials Needed:

- a pulley with a hook
- a pail with a handle
- a spring scale
- two pieces of rope (one short, one longer)
- two 1 lb. weights or a 2 lb. weight (500g weights or a 1 kg weight)

What To Do:

(Part 1)

1. Make a prediction about the answer to the question.

2. Fasten the short rope to the pail handle. Place the weight into the pail.

3. Attach the spring scale to the other end of the string. Lift the pail up using only the spring scale.

4. Record the effort needed to lift the pail as shown on the scale.

(Part 2)

5. Fasten the pulley to an object located above eye-level (coat rack, tree branch, playground equipment).

6. Place the longer string around the pulley and fasten one end to the pail handle. Place the weight inside the pail.

7. Attach the spring scale to the other end of the string. Lift the pail using only the scale by pulling down.

8. Record the effort needed to lift the pail as shown on the scale.

9. Make a conclusion about what you have observed.

Name: _____

Let's Predict

Does a pulley make lifting easier? _____

Let's Investigate!

Record the effort needed to lift the pail as shown on the scale for part one and part two of the experiment. Draw what happened during the experiment.

Drawings	Scale Reading
Part One	
Part Two	

Let's Conclude

Was your prediction correct? Explain.

_____

_____

_____

Do you think it is easier to pull up to lift an object or is it easier to pull down to lift an object? Explain.

_____

_____

_____

OTM-2173   ISBN: 978-1-4877-0201-4 © On The Mark Press

Block and Tackle Pulley System: Fixed and Movable

A block and tackle pulley system is made up of a fixed and a movable pulley. The fixed pulley allows you to pull down to lift a load, and the movable pulley lets you pull using only half of the effort. This is because the pulley rope takes half the weight of the load.

Question: How much less force is needed to lift a load using a block and tackle pulley system than using a simple fixed pulley system?

Materials Needed:

- 2 pulleys with hooks
- a pail with a handle
- a spring scale

- two long pieces of rope
- two 1 lb. weights or a 2 lb. weight (500 g weights or a 1 kg weight)

What To Do:

(Part 1)

1. Make a prediction about the answer to the question.

2. Create a fixed pulley system by fastening a simple pulley to an object located above eye-level (coat rack, tree branch, playground equipment).

3. Put a piece of long rope around the pulley and fasten one end to the pail handle. Place the weight inside the pail.

4. Attach the spring scale to the other end of the rope. Pull down on the spring scale to lift the pail. Record the effort needed to lift the pail.

(Part 2)

5. Create a block and tackle system. Keep the simple pulley fixed to the object located above eye-level. Tie the rope to the hook on the bottom of the fixed pulley. Thread the rope through a movable pulley, and back over top of the fixed pulley. Attach the spring scale to the rope.

6. Place the weight inside the pail. Hang the pail on the hook of the movable pulley.

7. Pull down on the spring scale to lift the pail. Record the effort needed to lift the pail as shown on the scale.

8. Make a conclusion about what you have observed.

Name: _____

Let's Predict:

How much less force is needed to lift a load using a block and tackle pulley system than using a simple fixed pulley system? (Use Newtons to explain.)

_____

_____

_____

Let's Investigate!

Record the effort needed to lift the pail as shown on the scale for part one and part two of the experiment. Draw what happened during the experiment.

Drawings	Scale Reading
Part One	
Part Two	

Let's Conclude

Was your prediction correct? Explain.

_____

_____

_____

_____

OTM-2173   ISBN: 978-1-4877-0201-4 © On The Mark Press

GEARING UP

LEARNING INTENTION:

Students will learn how gear trains operate to facilitate change in direction and speed. Students will construct devices that use wheels and axles to demonstrate and describe their use in gear systems.

SUCCESS CRITERIA:

- identify and make gear trains that facilitate change in direction and speed
- make predictions, observations, and record results of directionality and speed of gears using drawings and written descriptions
- make conclusions about how gear trains operate in order to change direction and speed

MATERIALS NEEDED:

- gears of different sizes (2 small, 2 medium, 2 large), per student/pair of students
- wooden boards or pieces of cardboard to use for bases, one per student/pair of students
- masking tape
- hammers, finishing nails
- safety goggles (a pair for each student)
- a copy of *Gearing Up!* Worksheets 1, 2, 3 and 4 for each student
- pencils, pencil crayons, markers

PROCEDURE:

1. Explain to students that they will create gear trains and investigate how they operate to facilitate change in direction and speed. Students can work individually or in pairs to conduct the investigation. Give students Worksheets 1, 2, and 3, and the materials they need. Read through the question, materials needed, and what to do sections on Worksheets 1 and 2, to ensure understanding. Students will conduct the experiment, record their observations for each part in the chart on Worksheet 2, and make conclusions on the directionality and speed of gears in a gear train through drawings on Worksheet 3.

2. Give students Worksheet 4, and the materials needed to take the *Gear Train Challenge!* Read through the instructions and allow students time to complete their predictions. They can then begin to work through the activity.

(Students should come to the conclusion that a pattern is created where the first and third gears in a gear train turn in the same direction, as does the second and fourth gear. When a fifth gear is added, it turns in the same direction as the first and third gear on the gear train.)

DIFFERENTIATION:

Slower learners may benefit by working together in a small group, with teacher direction, in order to complete the investigation. This would allow for small group instruction on how to manipulate the gears in order to facilitate change in direction and speed, and how to accurately measure and record observations.

For enrichment, faster learners could continue to expand their gear trains in order to create the longest one they possibly can. An alternate enrichment opportunity would be to access the internet to find photos of real-life machines that use gears. Photos could be printed out and used to create a collage. Students could then orally present collages to a small group to demonstrate where gears are used.

Name: _____

Gears

Gears are wheels with teeth. When the teeth on gears are connected, they move each other. This is called a gear train. Gearing up is when a large gear turns a smaller gear. Gearing down is when a small gear turns a larger gear. Idling is when two gears are turning at the same speed. Let's investigate the direction and speed at which gear trains turn!

Materials Needed:

- gears of different sizes (2 small, 2 medium, 2 large)
- finishing nails
- a hammer
- a marker
- a wooden board, or cardboard for the base
- masking tape
- safety goggles

What To Do:

(Part 1)

*Using a piece of tape and a marker, mark one tooth on each gear to help you count the number of times the gear turns.

1. Attach one small gear to the wooden board or cardboard base by hammering a nail in the centre of the gear. Attach another small gear next to it so that the teeth interlock.

2. Turn the first gear. Use these questions to guide your investigation:

   - *What direction does the second gear turn?*
   - *If you turn the first gear once around, how far does the second gear turn?*
   - *Is this gear train gearing up, gearing down, or idling?*

3. Record your observations in the chart on Worksheet 2.

OTM-2173   ISBN: 978-1-4877-0201-4 © On The Mark Press

Name: _____

(Part 2)

5. Replace the second small gear with a medium sized gear.

6. Turn the small gear. Use these questions to guide your investigation:

   • *What direction does the medium gear turn?*
   • *If you turn the small gear four full turns, how many times does the medium gear turn?*
   • *Is this gear train gearing up, gearing down, or idling?*

7. Record your observations in the chart below.

(Part 3)

8. Replace the small gear with a large gear.

9. Turn the large gear. Use these questions to guide your investigation:

   • *What direction does the medium gear turn?*
   • *If you turn the large gear four full turns, how many times does the medium gear turn?*
   • *Is this gear train gearing up, gearing down, or idling?*

10. Record your observations in the chart.

First Gear	Direction it turns	Number of turns	Second Gear	Direction it turns	Number of turns	Gearing up? Gearing Down? Idling?
Small			Small			
Small			Medium			
Large			Medium			

Name: _____

Let's Conclude

What did you discover? Speed and direction that gears turn on a gear train! Draw a picture of each gear train you created.

Gears Idling

Gearing Up

Gearing Down

OTM-2173   ISBN: 978-1-4877-0201-4 © On The Mark Press

Name: _____

Challenge!

Add one more gear to your already made gear train so that you have meshed three gears. What direction will the gears on your gear train turn?

I predict:

_____

_____

_____

Turn the first gear. Draw your observations. Use arrows to indicate the direction each gear is turning.

Add another gear to your gear train. Turn the first gear. Use arrows to indicate the direction each gear is turning.

What direction would a fifth gear turn if it was added to your gear train?

_____

_____

GEARS IN MOTION

LEARNING INTENTION:

Students will explain the operation of a drive system that uses a belt, chain and gears. Students will explain the operation of a drive system that transfers motion from one shaft to a second shaft either parallel or perpendicular (90 degrees) to the first.

SUCCESS CRITERIA:

- identify bevel gears and worm drive gears as facilitators to make motion happen
- conduct experiments to observe how gears create motion in everyday objects
- make and record observations using drawings and written descriptions
- make conclusions about how gears operate in order to change direction and speed

MATERIALS NEEDED:

- bevel gears and worm drive gears (to use as manipulatives)
- hand held egg beater/mixer (one or more if students do experiment in small groups)
- an electric mixer (opened to allow mechanical parts to be seen)
- a bicycle with multiple gear settings
- a piece of chalk
- clipboards (one per student)
- a copy of *Gears In Motion* Worksheets 1 and 2 for each student
- a copy of *Exploring the Bevel Gear* Worksheet 3 for each student
- a copy of *Exploring the Worm Gear Drive* Worksheet 4 for each student
- a copy of *How Bicycles Work* Worksheets 5, 6, and 7 for each student
- pencils, markers

PROCEDURE:

*This lesson could be done as one long lesson or divided into shorter lessons.

1. Give students Worksheets 1 and 2, in order to read through the information about bevel gears and worm gear drives. Some teacher led discussion about key concepts may be beneficial to help students understand the function of these gears, partnered with using actual bevel gears and worm gear drives as manipulatives.

2. Explain to students that they are going to explore how bevel gears work to transmit power and work to make motion happen. For the bevel gear exploration, students could work in a large teacher directed group, or in small groups. Give students Worksheet 3, and the materials they need. Read through with them the section on what to do, to ensure understanding. Students will conduct the exploration and record their observations through drawings and written description.

   Optional: the Alberta Distance Learning Centre created a video to show an egg beater bevel gear turning: ***https://youtu.be/iVJcRTqxztg***

3. This is a teacher directed exploration. Come together as a large group. Explain to students that they are going to explore how a worm gear drive works to transmit power and works to make motion happen. Give students Worksheet 4 and a clipboard. Read through with them the section on what to do, to ensure understanding. The electric mixer must be opened up so that its mechanical parts are visible. The teacher will operate the mixer to demonstrate how the worm gear drive works to transmit power to make the paddles turn. (Having students approach in small numbers is recommended so that they can get a closer look at the demonstration). Students record their observations through drawings and written description.

4. Explain to students that they are going to explore the gears on a bicycle to see how they work to make the bike move. Place the bicycle upside

OTM-2173   ISBN: 978-1-4877-0201-4 © On The Mark Press

down so that it is resting on its seat and handle bars. Give students Worksheets 5, 6, and 7, and a clipboard to use as they record their observations during this large group exploration. Read through the information and what to do sections on the worksheets, as the students complete each step. They will record their observations in the chart on Worksheet 7, and make conclusions based on their observations.

DIFFERENTIATION:

Slower learners may benefit by working with a peer to discuss ideas about their conclusions on how bicycles work. Scribing of responses to conclusions may also be beneficial.

For enrichment, faster learners could research other parts on a bicycle (i.e., brake system), in order to learn how all the parts work together to make the bicycle operate.

OTM-2173   ISBN: 978-1-4877-0201-4 © On The Mark Press

Name: _____

Gears In Motion

There are many types of gears that work to change speed and direction of motion. Some are designed to keep the speed steady, and some are designed to reduce speed. Gears can be connected in many ways in order to work to make motion happen.

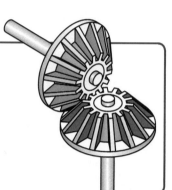

Bevel Gears

Bevel gears are most often attached to shafts that are 90 degrees apart. Bevel gears are found where the axes of the two shafts meet. The tooth-bearing faces of the gears are shaped like a cone. This allows the gears to go around corners.

Where could bevel gears be found?

Bevel gear on the inside of a watch.

Bevel gear lifts a floodgate by means of a central screw.

Bevel gear on an antique hand drill.

OTM-2173   ISBN: 978-1-4877-0201-4 © On The Mark Press

Name: _____

Worm Drive

A worm drive is a gear system where a worm (which is a gear in the form of a screw) meshes with a worm gear (which looks like a spur gear). A worm gear system transmits power from the worm to the gear wheel. The worm acts like a brake. The worm gear needs to rotate many times for each complete turn of the gear wheel.

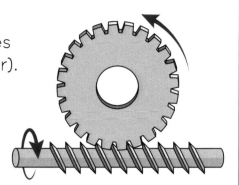

Where could worm drives be found?

A double base features worm gears as tuning mechanisms.

This is a worm drive controlling a gate.

OTM-2173   ISBN: 978-1-4877-0201-4 © On The Mark Press

Name: _____

Exploring the Bevel Gear

Let's explore how bevel gears work to transmit power
and work to make motion happen!

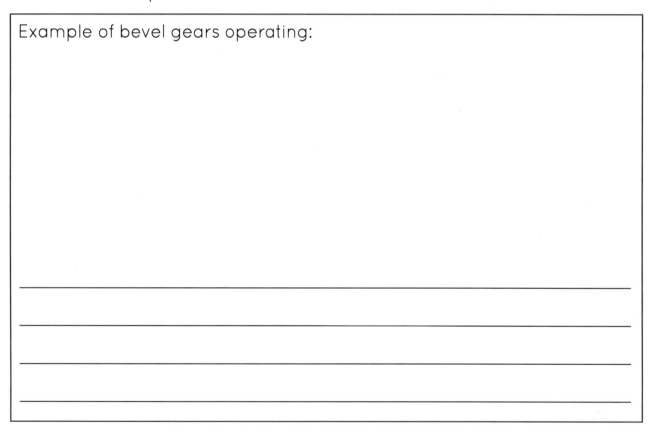

Materials Needed:

* a hand held egg beater

What To Do:

1. Turn the egg beater handle. Carefully observe how the bevel gears
    connect in order to make the beater paddles turn.

2. Record your observations by drawing what you see happening. Write
    a sentence to describe how the gears in the egg beater work to create
    motion of the paddles.

Example of bevel gears operating:

_____

_____

_____

_____

You turned the handle up and down but the beaters turned around and
around. Are the gears operating at 90 degrees apart from each other?
Or do they operate parallel to each other?

_____

  OTM-2173   ISBN: 978-1-4877-0201-4 © On The Mark Press

Name: _____

Exploring the Worm Gear Drive

Let's explore how a worm gear drive works to transmit power and works to make motion happen!

Materials Needed:

- an electric mixer (opened to allow the inside to be seen)

What To Do:

1. Your teacher will open up the electric mixer so that the parts on the inside are visible. Your teacher will plug in the mixer and operate it so that you are able to view and carefully observe how the worm gear drive works to create motion of the paddles.

2. Record your observations by drawing what you see happening. Write a sentence to describe how the gears in the electric mixer work to create motion of the paddles.

Example of a worm drive operating:

_____

_____

_____

Name: _____

How Bicycles Work

We have learned a lot about how gear trains and pulleys systems work to make things move. Let's explore the parts of a bicycle to see them in motion!

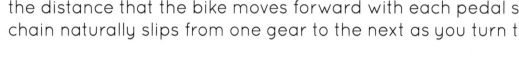

The chain of a bicycle is like a belt drive pulley system. The pedals are like an input wheel and the back gears are like an output wheel, always turning in the same direction around the sprockets, as the gears are changed. The idea behind multiple gears on a bicycle is to let you change the distance that the bike moves forward with each pedal stroke. The chain naturally slips from one gear to the next as you turn the pedals.

Materials Needed:

- a bicycle with multiple gears settings
- a piece of chalk

What To Do:

(Part A)

1. Turn the bicycle upside down so that it is resting on the handle bars and seat.

2. Using the chalk, make a mark somewhere on the back tire.

OTM-2173   ISBN: 978-1-4877-0201-4 © On The Mark Press

Name: _____

(Part B)

3. Put the bicycle in low gear. Count the number of gear teeth on the front and rear sprockets that are being used. Record your observations in the chart on Worksheet 7.

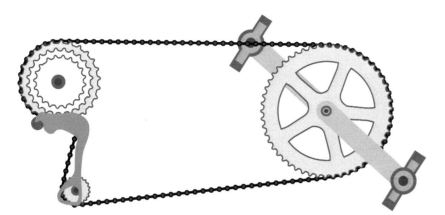

The bicycle is in low gear when the chain passes around the largest rear sprocket.

4. Turn the pedal on the bicycle one complete turn. Observe how many turns the back wheel makes. Record your observations in the chart.

(Part C)

5. Put the bicycle in high gear. Count the number of gear teeth on the front and rear sprockets that are being used. Record your observations in the chart.

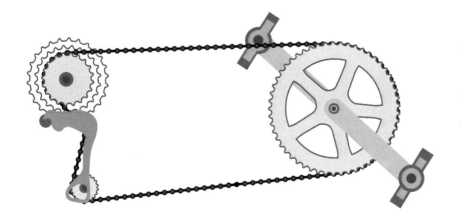

The bicycle is in high gear when the chain passes around the small rear sprocket.

6. Turn the pedal on the bicycle one complete turn. Observe how many turns the back wheel makes. Record your observations in the chart.

7. Make conclusions about what you observed.

Name: _____

Let's Observe

	Number of gear teeth on front sprocket	Number of gear teeth on rear sprocket	Total Number of gear teeth being used	Number of turns rear wheel does with one pedal turn
Low Gear				
High Gear				

Let's Conclude

Describe the force you needed to use in order to turn the pedal when putting the bicycle in low gear.

_____

_____

_____

Describe the force you needed to use in order to turn the pedal when putting the bicycle in high gear.

_____

_____

_____

Which gear would be best to use when travelling up a steep hill? Which gear would be best to use when traveling fast on flat ground? Explain your thinking.

_____

_____

_____

You turn the pedals of a bicycle up and down and the tires turn up and down as well. Are the gears and wheels operating at 90 degrees apart from each other? Or do they operate parallel to each other?

_____

OTM-2173   ISBN: 978-1-4877-0201-4 © On The Mark Press

WHEELS IN MOTION

LEARNING INTENTION:

Students will explain how wheels and rollers can be used to move an object and demonstrate the use of rollers in a practical way. Students will compare the wheel and the roller and identify examples where each are used. Students will construct and explain the operation of a drive system and construct devices that use wheels and axles in model vehicles.

SUCCESS CRITERIA:

- identify objects that use wheels in order to move things
- make a prediction, investigate how wheels/ rollers work to move an object, make observations and record effort ratings using a spring scale
- design and make a plan to construct a car, record outcomes of the final product by testing for movement of cars using muscular force, gravitational force, wind energy force, or mechanical force
- make conclusions about the advantages in the use of a wheel
- make connections to people and the environment

MATERIALS NEEDED:

- shoe boxes with lids (one per group of students)
- ping pong balls or other balls of about that size (4 per group of students)
- blocks of wood, pre-cut to measure 5 cm × 10 cm × 15 cm (2 in. × 4 in. × 6 in.) (one per group of students)
- spring scales (one per group of students)
- extra pieces of cardboard, duct tape, scissors, string, glue, construction paper, markers, butterfly fasteners, hole puncher, straws, elastics, pipe cleaners, aluminum foil, cardboard tubes, masking tape (or any other materials that you find suitable to include)
- short pieces of dowel for axles in cars (two pieces for each student)

- a copy of **Wheels In Motion** Worksheet 1 for each student
- a copy of **Let's Get Rolling** Worksheets 2 and 3 for each student
- a copy of **Constructing a Car** Worksheets 4 and 5 for each student
- a copy of **The Turbo Challenge** Worksheets 6 and 7 for each student
- pencils

PROCEDURE:

*This lesson could be done as one long lesson, or divided into shorter lessons.

1. Students will take a look at a simple machine called a wheel and axle. Give students Worksheet 1, and read through the description of a wheel with them. They will colour the wheels on the objects, then draw and label other objects that use wheels to move.

2. Divide students into small groups, and give them the materials needed to conduct the investigation about the use of wheels. Read through the question, materials needed, and what to do sections on Worksheet 2. Students will make a prediction about the outcome of the investigation, make observations through drawings, record the force of effort used to move the block of wood, with and without the use of wheels, then make a conclusion on Worksheet 3. A large group discussion as to how the wheels/rollers could operate better to facilitate movement would be beneficial. Discussing how a wheel and axle system works will aid students in the car building activity to follow.

3. Students will plan, design, and construct a car. They will compare their final product with a partner and make some observations. Give students a copy of the **Constructing a Car** Worksheets 4 and 5 to complete, and the materials to construct their cars (shoeboxes can be given to groups to use for the body of their car. Upon completion, students can present their cars to the class and share three things about them.

4. Students will plan, design, and enhance their car's movement. Give each student a copy of **The Turbo Challenge!** Worksheets 6 and 7. Read through with the students, the section regarding the force enhancement choices needed to be made in order to turbo charge their cars. Students will make a plan and draw a design of their turbo charged cars before they begin construction. Once their invention is completed, students will work with a partner to give a demonstration of how their car was enhanced and how it performs to incorporate a force causing movement. Students will record input given from their partner (a star = something done well, a wish = something that needs to be improved).

DIFFERENTIATION:

Slower learners may benefit by listing only one thing that is the same and one thing that is different from their partner's car on Worksheet 5, and provide only one thing to share with the class about their cars. They may also benefit by working with a partner to plan and design their turbo enhanced cars. The implementation of the plan could then be completed as a team or independently.

For enrichment, faster learners could discuss in a small group, where they have seen wheels or rollers used and how they were helpful in reducing the work effort needed to move objects.

OTM-2173   ISBN: 978-1-4877-0201-4 © On The Mark Press

Name: _____

Wheels in Motion

The invention of the wheel is considered to be one of the most important discoveries of all time. It is made up of a round wheel which turns around an inner cylinder called an axle. This simple machine allows an object to roll. We can find wheels on many machines such as lawn mowers and carts, and on vehicles such as cars, trains, planes, and bicycles.

Colour the wheels on each object below.

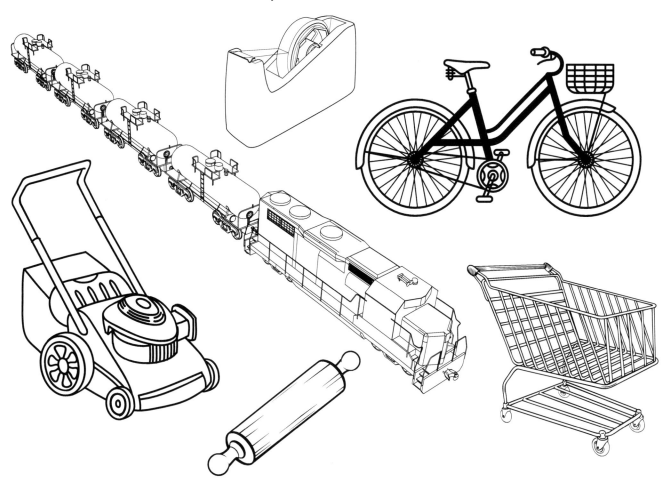

Draw and label three other objects that use wheels to move.

Name: _____

Let's Get Rolling!

Question: Does a wheel or roller make moving the objects easier?

Materials Needed:

- a shoe box lid, and pieces of extra cardboard
- 4 ping pong balls or other balls of about that size
- a block of wood 5 cm × 10 cm × 15 cm

- duct tape
- scissors
- string 20 cm in length
- a spring scale

What To Do:

(Part 1)

1. Attach the string to the top of the block of wood using the duct tape.

2. Attach the other end of the string to the spring scale.

3. Using the spring scale handle, pull to drag the block of wood across a flat surface. Observe and record the force of effort needed to move the block of wood in the chart on Worksheet 3.

(Part 2)

4. Using extra cardboard as material, cut out 2 pieces of cardboard long enough to span the width of the shoe box lid.

5. Attach these pieces to the underside of the shoebox lid using duct tape. Ensure they span the width of the box lid, and are placed about 7 cm from the ends of the lid.

6. Attach the block of wood (with string and spring scale still attached) to the top of the shoebox lid using duct tape.

7. With your partner's help, place the shoebox lid on top of the 4 ping pong balls (ensure that 2 balls are placed at each end of the box lid).

8. Using the spring scale handle, pull to drag the block of wood across a flat surface. Observe and record the force of effort needed to move the block of wood now, in the chart on Worksheet 3.

9. Make a conclusion about what you observed.

OTM-2173   ISBN: 978-1-4877-0201-4 © On The Mark Press

Name: _____

Let's Predict

Does a wheel or roller make the moving of objects easier?

_____

Let's Investigate

Record the force of effort needed to move the block of wood across a flat surface. Draw what happened during the experiment.

Drawings	Force of Effort Used (in Newtons)
Part One	
Part Two	

Let's Conclude

Was your prediction correct? Explain.

_____

_____

_____

Name: _____

Constructing A Car

Use what you know about the advantages of the wheel and axle to design and construct a car.

The materials I will use are:

_____

_____

_____

My plan for making my car is:

1. _____

2. _____

3. _____

4. _____

5. _____

6. _____

Draw a design of what your car will look like:

Now carry out your plan by gathering the materials you need. Begin your construction!

OTM-2173   ISBN: 978-1-4877-0201-4 © On The Mark Press

Name: _____

Share and Compare

Working with a partner, compare your final product. List two things that are the same, and two things that are different about your cars.

Things That Are The Same	Things That Are Different
1. _____ _____ _____	1. _____ _____ _____
2. _____ _____ _____	2. _____ _____ _____

Three things I would like to share with the class about my car are:

1. _____

_____

2. _____

_____

3. _____

_____

What might you change about your car?

_____

_____

Name: _____

The Turbo Challenge!

How powerful can your car become with a little **turbo** added to it?
Choose a simple force to power up your car, and get started!

Choose your force enhancer:

- **Muscular force:** direct push or pull
- **Gravitational force:** downhill motion
- **Wind energy force:** wind, moving air
- **Mechanical force:** cranking devices, springs, elastic band propulsion

The materials I will use are:

_____

_____

_____

_____

_____

Draw a design of your turbo charged car:

My plan for enhancing the force on my car is:

1. _____

2. _____

3. _____

4. _____

5. _____

6. _____

Carry out your plan by gathering the materials you need. Begin creating your turbo charged car!

OTM-2173   ISBN: 978-1-4877-0201-4 © On The Mark Press

Name: _____

Let's Test It

Working with a partner, give a demonstration to show how you enhanced the force on your car and how it performs. Describe what happened.

_____

_____

_____

Thoughts from my partner about my turbo charged car...

A star:

_____

_____

_____

A wish:

_____

_____

Improve The Move!

Use a drawing to show the design changes needed to improve the turbo charge on your car.

LEVERS

LEARNING INTENTION:

Students will demonstrate ways to use a lever that applies a small force to create a large force and apply a small movement to create a large movement. Students will predict how changes in the size of a lever or the position of the fulcrum will affect the forces and movement to create a larger movement. Students will construct models of levers and explain how levers are involved in objects in their local environment.

SUCCESS CRITERIA:

- identify and label the parts of different types of levers (effort, load, fulcrum)
- investigate how levers help to move or lift heavy things
- make observations by recording effort ratings on a scale and completing diagrams
- make conclusions about the use of levers
- make connections to people and items in the environment

MATERIALS NEEDED:

- a wooden board about 4 in. (120 cm) long
- a small wooden block
- several textbooks
- tweezers, a coin (a set per group of students)
- examples of first, second, third class levers to use as manipulatives (e.g., nutcracker, stapler, rake, tweezers, pliers, tennis racket, scissors, baseball bat, scale, wheelbarrow)
- a copy of *Levers* Worksheets 1, 2, and 3 for each student
- a copy of *Looking at Levers* Worksheet 4 for each student
- a copy of *Lever Leverage* Worksheets 5, 6, and 7 for each student
- pencils

PROCEDURE:

***This lesson could be done as one long lesson or divided into shorter lessons.**

1. Students will take a closer look at a simple machine called a lever. Give them Worksheet 1 to complete. Read through the description of a lever, using the picture to explain where the effort, load, and fulcrum are located on the lever.

2. Students will experiment with levers. Give them Worksheets 2 and 3. Read through the question, materials needed, and what to do sections for the experiment. Conduct the investigation as a large group, ensuring that each student has an opportunity to experiment with the effort needed to lift the load as the fulcrum is moved on the lever. Students will rate the effort needed to lift the load on each lever, and make a conclusion on how a lever works to move or lift things.

3. Give students Worksheet 4. Read through and discuss the information with them about the different types of levers. Having a sample of each type of lever to use as manipulatives throughout the discussion may be beneficial. Students will experiment with a third class lever. Give them Worksheets 5 and 6, and the materials to conduct the investigation. Read through the question, materials needed, and what to do sections for the experiment. Students will conduct the investigation, rate the effort needed to lift the coin each time and draw a diagram to show their efforts, then make a conclusion about their findings. Give students Worksheet 7. They will identify where the effort, load, and fulcrum are located on some everyday objects.

DIFFERENTIATION:

Slower learners may benefit by working with a partner and by having access to actual objects that are depicted on worksheet 7 (as manipulatives) while completing the activity to identify the effort, load, and fulcrum.

For enrichment, faster learners could locate a few objects in the classroom and determine if the objects are levers by testing if they lift or move something; and identify where the effort, load, and fulcrum are located on the objects.

OTM-2173   ISBN: 978-1-4877-0201-4 © On The Mark Press

Name: _____

Levers

A lever is a simple machine that helps you to move or lift a weight with little effort. It is a bar that turns on a point called a fulcrum. There are three important parts on a lever. They are effort, fulcrum, and load.

Draw a picture of you and a friend on a seesaw. Write the words load, effort, and fulcrum where they belong on your picture.

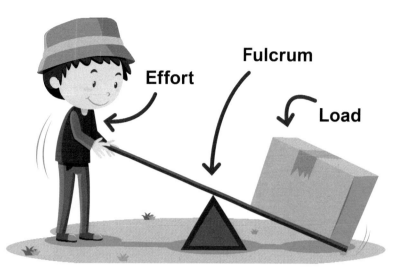

Effort **Fulcrum** **Load**

OTM-2173   ISBN: 978-1-4877-0201-4 © On The Mark Press

Name: _____

Question: Do levers help to move or lift heavy things?

Materials Needed:

- a wooden board about 120 cm long (lever)
- a small block of wood (fulcrum)
- several textbooks (load)

What To Do:

1. Place the wooden block on a flat, level surface.

2. Place the board on the wooden block so that the board is balanced.

3. Place the textbooks on one end of the board.

4. Using your hand, press down the other end of the board to lift the textbooks.

5. Describe the effort needed to lift the books by putting a mark on the effort meter on Worksheet 3.

(1 = least effort needed, 5 = greatest effort needed)

6. Move the fulcrum in a direction away from the textbooks. Repeat steps 4 and 5.

7. Move the fulcrum in a direction toward the textbooks. Repeat steps 4 and 5 again.

8. Make a conclusion about what you observed.

OTM-2173 ISBN: 978-1-4877-0201-4 © On The Mark Press

Name: _____

Levers Investigation Sheet

Investigation	Effort Meter
Diagram One	**5** – Greatest Effort **4** – **3** – **2** – **1** – Least Effort
Diagram Two	**5** – Greatest Effort **4** – **3** – **2** – **1** – Least Effort
Diagram three	**5** – Greatest Effort **4** – **3** – **2** – **1** – Least Effort

Let's Conclude

Fill in the blanks with the words from the word box below.

> **less   effort   closer**

The _____ needed to lift the textbooks is

_____ when the fulcrum is _____

to the load.

Name: _____

Looking At Levers

In the previous investigation, we looked at how a **first class lever** best performs to move or lift a load. There are other types of levers, and the fulcrum does not always have to be located between the effort and the load. The **effort**, **load**, and **fulcrum** can be arranged in any combination along the bar.

First Class Lever

In this lever, the fulcrum is located between the effort and the load.

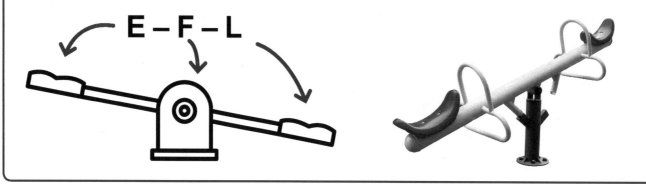

E – F – L

Second Class Lever

In a second class lever, the load is located between the effort and the fulcrum.

F – L – E

Third Class Lever

In a third class lever, the effort is located between the load and the fulcrum.

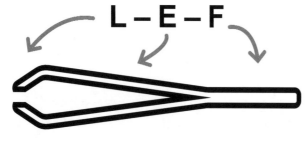

L – E – F

OTM-2173   ISBN: 978-1-4877-0201-4 © On The Mark Press

Name: _____

Lever Leverage

Question: Does changing the location of the effort on a third class lever affect its performance?

Materials Needed:

- a pair of tweezers
- a coin

What To Do:

(Part 1)

1. Place your thumb and index finger on the tweezers close to the fulcrum.

2. Squeeze the tweezers to pick up the coin. Describe the effort needed to lift the coin by putting a mark on the effort meter on Worksheet 6.

(1 = least effort needed, 5 = greatest effort needed).

(Part Two)

3. Move your fingers to a location on the tweezers midway between the fulcrum and the load.

4. Repeat step 2.

(Part Three)

5. Move your fingers closer to the load.

6. Repeat step 2.

7. Make a conclusion about what you observed.

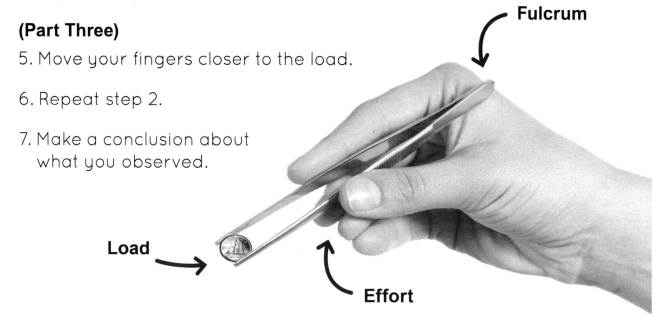

Fulcrum

Load

Effort

Name: _____

Levers Leverage! Investigation Sheet

Diagram of the tweezers and the coin	Effort Meter
Part One: label the fulcrum, load, and position of the effort	**5** – Greatest Effort **4** – **3** – **2** – **1** – Least Effort
Part Two: label the fulcrum, load, and position of the effort	**5** – Greatest Effort **4** – **3** – **2** – **1** – Least Effort
Part Three: label the fulcrum, load, and position of the effort	**5** – Greatest Effort **4** – **3** – **2** – **1** – Least Effort

Let's Conclude

Fill in the blanks with the words from the word box below.

> **less   effort   closer**

The _____ needed to lift the coin is considerably

_____ when it is applied _____

to the load.

OTM-2173   ISBN: 978-1-4877-0201-4 © On The Mark Press

Name: _____

Lever Leverage! Apply the Knowledge!

Label where the **effort** (E), **load** (L), and **fulcrum** (F) are found on the objects below.

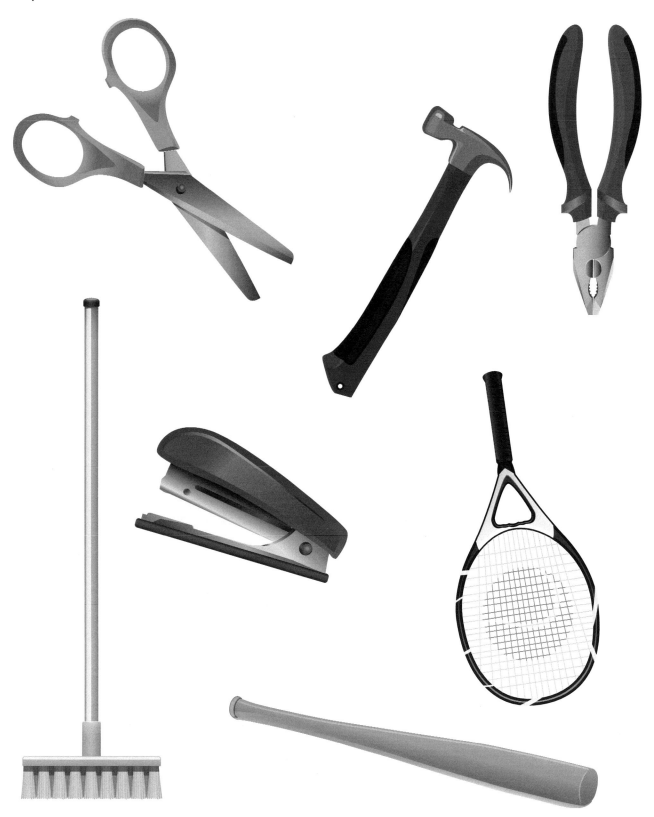

OTM-2173   ISBN: 978-1-4877-0201-4 © On The Mark Press

TOPIC C: BUILDING DEVICES AND VEHICLES THAT MOVE: POWERED UP VEHICLES

LEARNING INTENTION:

Students will design and construct vehicles that move and have moving parts. Students will use simple forces to power or propel a device. Students will design and construct vehicles that employ energy-storing components that will cause motion. Students will compare two designs, identifying the relative strengths and weaknesses of each

SUCCESS CRITERIA:

- read, follow and review steps in a process to complete the construction of a vehicle project
- test a vehicle using criteria, controls and record results of tests
- make observations, conclusions, and connections to procedures and projects with opportunities to evaluate and improve designs

MATERIALS NEEDED:

- a copy of *Balloon Power!* Worksheets 1 and 2 for each student
- a copy of *Rubber Band Power!* Worksheets 3 and 4 for each student
- a copy of *Three Challenges: Speed, Distance and Creativity* Worksheets 5, 6 and 7 for each student
- sheets of cardboard or strong boxboard or carton material cut to rectangles 10 cm by 20 cm
- straws (preferably thin), skewers, toothpicks, bottle tops, nails
- balloons
- rubber bands,
- tape – masking, clear or duct
- hot glue guns, or similar joining material
- scissors
- plasticine or modelling clay (optional, for weight and stability in wheels)
- measuring tape

PROCEDURE:

***This lesson can be done in one long lesson or divided into shorter lessons.**

1. Lead students in a discussion on what makes vehicles move. Introduce the ideas of force, energy, power, propulsion and fuel. Give students Worksheets 1 and 2. Read through the worksheets and review steps to make sure students understand what to do. Prepare a model vehicle ahead of time and demonstrate a distance trial. Ask students to think about if the vehicle went in a straight line or turned one way or the other. Monitor the work of students and review the work when everyone is finished with trials.

   This project is based on the video: **https://youtu.be/IacekOC-gwI**. An internet search for "balloon powered car" will offer variations, such as using a water bottle for the vehicle body.

2. Explain that last time the students used a ballon and air to propel a device. Another way to move a vehicle would be the stored energy in a wound up rubber band. Give students Worksheets 3 and 4. Read through the worksheets and review steps to make sure students understand what to do. Prepare a model vehicle ahead of time and demonstrate a distance trial. Ask students to think about if the vehicle went in a straight line or turned one way or the other. Monitor the work of students and review the work when everyone is finished with trials.

   This car is based on the video: **https://youtu.be/hjXYVpAyYG0**. An internet search for "rubber band powered car" will offer variations.

3. Lead students in a discussion on different types and ways to power vehicles. The class used balloons and rubber bands. Could a spring power one of their cars? How about a motor or fan or a sail? Tell students they will have a chance to apply what they have learned from the two

OTM-2173 ISBN: 978-1-4877-0201-4 © On The Mark Press

vehicles they have made. But this time, they face three challenges – speed, creativity and distance. Give students Worksheets 5 and 6. Read through the instructions and review the material for understanding. Prepare the testing grounds for the students. For the speed challenge, each vehicle will be timed over a distance of 3 metres (distance as space allows). For the creativity challenge, students will display their vehicles and receive ratings from other students. Students can use a rating scale of 1 to 5 stars (see Worksheet 7 for printable rating ballots). For the distance challenge, measure the total distance a vehicle moves in one run (the class may need to use a hallway or open space).

DIFFERENTIATION:

For slower learners or students who have trouble with handling objects, prepare materials with the holes or cuts to accommodate the work to their pace and manual dexterity. Pair up students for construction tasks as needed.

For enrichment, faster learners could experiment with different construction techniques to improve speed or distance travelled by vehicles - larger or smaller wheels, heavier or lighter car bodies and different balloon types.

Name: _____

Balloon Power!

Question: How can you power a toy car with a balloon?

What You Need:

- 4 bottle tops
- wooden skewers
- straws – cut to 10 cm lengths
- a balloon
- a thin nail
- a piece of cardboard or carton material cut to a rectangle for the vehicle body – about 10 cm by 20 cm
- tape, scissors
- a measuring tape or distances marked on the ground

What To Do:

1. Make a small hole in the centre of each bottle top using the nail.

2. Slide one skewer into one wheel. Slide a straw over the skewer. Attach another wheel to the other end of the skewer. This will be your axle. Make sure the wheels and skewers do not fall apart.

4. Repeat steps 1 and 2 for the other set of wheels.

5. Tape the straws to the vehicle body so they line up like the wheels on a car.

6. Tape the balloon opening around another piece of straw. This will be your "motor". Make sure the tape is sealed tight so that you can use the straw to blow up the balloon.

7. Tape your motor to the top of your vehicle so that the straw sticks out 3cm from the back end.

8. Blow up the balloon. Pinch the straw to keep the air from getting out. Now test your vehicle! Do at least 3 trials to see how far it goes.

OTM-2173   ISBN: 978-1-4877-0201-4 © On The Mark Press

Name: _____

Let's Observe

Draw a picture of your balloon powered vehicle. Label each part.

How far did your vehicle go?

	Distance
Trial 1	
Trial 2	
Trial 3	

Think About It!

Did it travel in a straight line or did it turn while moving?

_____

_____

What could you do to the design to help your vehicle move in a straight line?

_____

_____

_____

Name: _____

Rubber Band Power

Question: How can you power a toy car with a rubber band?

What You Need:

- four bottle tops
- wooden skewers
- toothpick, cut to 0.5 cm in length
- straws – one cut to 10 cm in length, two cut to 5 cm in lengths
- rubber bands

- a thin nail
- piece of cardboard or carton material, cut to about 10 cm by 20 cm
- hot glue or tape, scissors
- a measuring tape or distances marked on the ground

What To Do:

1. Make a small hole in the centre of each bottle top using the nail.

2. Slide one skewer into one wheel. Slide a 10 cm long straw over the skewer. Attach another wheel to the other end of the skewer. This will be your front axle. Make sure the wheels and skewers do not fall apart. Slide a rubber band over the straw.

3. Slide one skewer into one wheel. Slide two 5 cm long straws over the skewer. Attach another wheel to the other end of the skewer. This will be your back axle. Make sure the wheels and skewers do not fall apart.

4. Glue a short piece of toothpick to the very centre of your back axle and between the two pieces of straw. This is going to be a hook for the rubber band and it will be an important part of your motor.

5. On the 10 cm side of your piece of cardboard, use a ruler to find the centre. Cut a notch or gap out of the cardboard on that side about 2 cm wide and 2 cm long. Do the same on the other 10 cm side of your piece of cardboard. This will be the body of the car. The notches you cut out of the cardboard will be a gap for the rubber band.

6. Tape the pieces of straw on the back axle to the cardboard. The rubber band on the front axle should be in the gap you made in step 5. The short piece of toothpick on the rear axle should stick out in the gap on the back end of your vehicle.

7. Stretch your rubber band so that it goes around the piece of toothpick.

OTM-2173   ISBN: 978-1-4877-0201-4 © On The Mark Press

8. Turn your back wheels so that the rubber band rolls up around your back axle several times. Now test your vehicle! Do least 3 trials to see how far it goes.

Let's Observe

Draw a picture of your rubber band powered vehicle. Label each part.

How far did your vehicle go?

	Distance
Trial 1	
Trial 2	
Trial 3	

Think About It!

Did it travel in a straight line or did it turn while moving?

_____

_____

What could you do to the design to help your car move in a straight line?

_____

_____

_____

OTM-2173   ISBN: 978-1-4877-0201-4 © On The Mark Press

Name: _____

Three Challenges: Speed, Distance and Creativity

You made a car propelled by a balloon and by a rubber band. Now you need to make a car that will travel fast, go far and look amazing! The vehicle must be smaller than 15 cm by 25 cm and use bottle tops for wheels like in the last projects.

The materials I will use are:

_____ _____

_____ _____

_____ _____

_____ _____

_____ _____

My plan for making the vehicle is:

1. _____

2. _____

3. _____

4. _____

5. _____

6. _____

We used rubber bands and balloons to propel other vehicles. My vehicle

will be propelled by _____

Explain why you chose this kind of propulsion:

_____

_____

OTM-2173   ISBN: 978-1-4877-0201-4 © On The Mark Press

Name: _____

Draw a design of what your vehicle will look like.

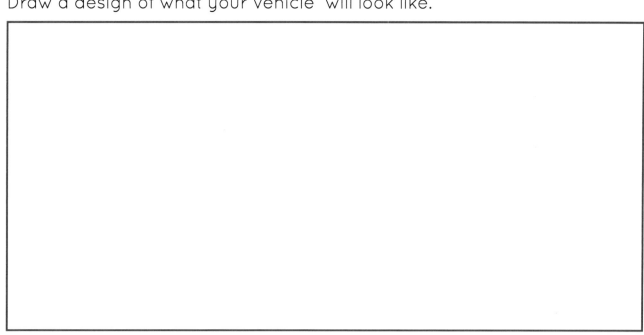

Carry out your plan by gathering the materials you need and create your new vehicle!

Three Challenges

Speed

It took my vehicle _____

seconds to go _____

metres.

Distance

My vehicle went

_____ metres.

Creativity

I give my vehicle _____ stars because:

_____

_____

_____

_____

Name: _____

Creativity Challenge

Student Name: _____

Rating:

⭐1 ⭐2 ⭐3 ⭐4 ⭐5

Explanation:

_____

_____

_____

Creativity Challenge

Student Name: _____

Rating:

⭐1 ⭐2 ⭐3 ⭐4 ⭐5

Explanation:

_____

_____

_____

Creativity Challenge

Student Name: _____

Rating:

⭐1 ⭐2 ⭐3 ⭐4 ⭐5

Explanation:

_____

_____

_____

Creativity Challenge

Student Name: _____

Rating:

⭐1 ⭐2 ⭐3 ⭐4 ⭐5

Explanation:

_____

_____

_____

Creativity Challenge

Student Name: _____

Rating:

⭐1 ⭐2 ⭐3 ⭐4 ⭐5

Explanation:

_____

_____

_____

Creativity Challenge

Student Name: _____

Rating:

⭐1 ⭐2 ⭐3 ⭐4 ⭐5

Explanation:

_____

_____

_____

OTM-2173 ISBN: 978-1-4877-0201-4 © On The Mark Press

THE CATAPULT (CONTROL! CONTROL! YOU MUST LEARN CONTROL!)

LEARNING INTENTION:

Students will recognize the need for control in mechanical devices and apply control mechanisms where necessary. Students will identify steps to be used in constructing a device and work cooperatively with other students to construct the device. Students will design and construct several different models of a device and evaluate each model, working cooperatively with other students.

SUCCESS CRITERIA:

- investigate, propose and carry out methods for improving designs and projects
- collaborate with others in the design, step-by-step construction and testing of a device
- adopt an attitude of improvement and opportunity in the process of building devices

MATERIALS NEEDED

- a copy of *Making a Catapult* Worksheets 1, 2, 3 and 4 for each student
- popsicle sticks, or craft sticks, regular and larger craft sticks
- nail file or a rasp tool
- rubber bands
- eye protection goggles
- target buckets – sandcastle buckets or something similar is appropriate
- Launch ammunition – soft candy like jujubes, small foam balls or something similar would be appropriate and not cause damage to anything. A cylindrical or spherical shape with some mass would be ideal.
- plastic spoons, large bottle tops, string or twine (optional)
- tape, glue, hot glue and joining material
- measuring tape
- rulers, pencils

PROCEDURE:

1. Lead students on a discussion of aiming. If you want your friend to catch a ball you are going to throw, you have to aim so that it comes close to them. Do you move around a lot when you throw the ball? Do you look at your friend or look somewhere else? These kinds of things give you more control, and you can do the same thing with mechanical devices you build. Give students Worksheets 1 and 2. Read through the worksheets and review steps to make sure students understand what to do. Prepare a model vehicle ahead of time and demonstrate a target trial with a bucket. Monitor student progress and when students are ready set up a firing range. Be sure to review and use safety procedures.

    This project is based on the video: **https://youtu.be/XchdUB-ZnKc**. Other designs and ideas are available with an internet search.

2. Discuss with students how successful they were at launching and placing their shots in or near the target buckets. Lead a discussion on things that could go wrong and how they might be able to make changes to the design for better control. What if the catapult had a bucket for the ammunition? (Option: glue the large bottle top to the launching arm.) What if the catapult had a wider or stronger base? As a supplement, show students videos of catapult projects. Here are some examples:

    https://youtu.be/iKQaTFfhwWo
    https://youtu.be/5ghagun8BPc

    Discuss with students what kind of control mechanisms were used.

3. Give students Worksheets 3 and 4. Read through the instructions and check for understanding. Monitor their progress and when students are ready set up a firing range. Be sure to review and use safety procedures.

DIFFERENTIATION:

For slower learners or students who have trouble with handling objects, prepare materials with the holes or cuts to accommodate the work to their pace and manual dexterity. Pair up students for construction tasks as needed.

For enrichment, faster learners could test other kinds of material as launch ammunition. Examples might include a tape ball, a small foam ball or objects with different shapes. Before any test launches occur, be sure to review and use safety procedures.

OTM-2173   ISBN: 978-1-4877-0201-4 © On The Mark Press

Name: _____

Making a Catapult

Question: Can I build a catapult and use it to aim and hit a target?

Materials Needed:

- 8 regular sized popsicle sticks, or craft sticks
- 2 large craft sticks
- a nail file or a rasp tool
- 3 rubber bands
- eye protection goggles
- a bucket
- pencil, ruler

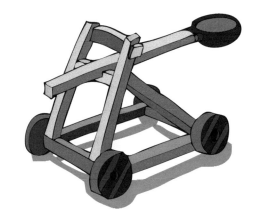

What To Do:

1. Draw a line across one of the large craft sticks 1 cm from one end. Do the same to the other large craft stick.

2. Use the nail file or rasp tool to make a groove on both ends of the lines you just drew on the large craft sticks.

3. Stack the 8 regular size sticks into a pile. Tie them together with a rubber band near each end. Make sure the bands are tight.

4. Take one large stick and push the end with the grooves between the last two regular size sticks in your pile.

5. Place the other large stick on top of your pile lined up so that the grooves match the grooves on the large stick on the bottom.

6. Use the grooves to tie the two large sticks together with a rubber band. Make sure this rubber band is more loose. You might only need to wrap it around 2 or 3 times.

7. Pull the large sticks back so that the grooves and the rubber band is close to the pile of other sticks. The top one should angle up a little bit.

Your catapult is ready. Now prepare for launch and aim for the bucket! Do at least 3 trials with your catapult. Be sure to measure how close you got if your launch didn't land in the bucket.

Name: _____

Let's Observe

Draw a picture of your catapult. Label each part.

[blank drawing box]

Did you land in the bucket? Measure how far you were from the target.

	Distance		
Trial 1	**Hit the Target?** Yes No	**Distance from Target** _____	
Trial 2	**Hit the Target?** Yes No	**Distance from Target** _____	
Trial 3	**Hit the Target?** Yes No	**Distance from Target** _____	

Let's Share

Find a partner and compare results. Who was more successful at hitting the target?

_____

_____

What could you and your partner do to get more control over your catapult launches?

_____

_____

_____

   OTM-2173   ISBN: 978-1-4877-0201-4 © On The Mark Press

Name: _____

Improving a Catapult

You made a catapult. How will you improve it? You and a partner will think about what you have learned, design a new catapult, build it and test it.

The materials we will use are:

_____   _____

_____   _____

_____   _____

_____   _____

Our plan for making the catapult is:

1. _____

2. _____

3. _____

4. _____

5. _____

6. _____

7. _____

8. _____

Your improvements to the catapult should give you more control over the catapult and more control over aiming it. Explain how the changes you have made to the design will give you more control.

_____

_____

_____

Name: _____

Draw a design of what your catapult will look like.

```
┌─────────────────────────────────────────────┐
│                                               │
│                                               │
│                                               │
│                                               │
│                                               │
│                                               │
│                                               │
│                                               │
│                                               │
└─────────────────────────────────────────────┘
```

Carry out your plan by gathering the materials you need and create your new catapult!

Let's Test it

Did you land in the bucket? If not, measure how far you were from the target.

	Distance		
Trial 1	**Hit the Target?** Yes No	**Distance from Target** _____	
Trial 2	**Hit the Target?** Yes No	**Distance from Target** _____	
Trial 3	**Hit the Target?** Yes No	**Distance from Target** _____	

Think About It

Were you more successful this time at hitting the target?   **Yes   No**

Why do you think this happened? _____

_____

If you made another catapult, what improvements would you make?

_____

_____

OTM-2173   ISBN: 978-1-4877-0201-4 © On The Mark Press

TOPIC D: LIGHT AND SHADOWS:
WHAT IS LIGHT?

LEARNING INTENTION:

Students will identify a range of sources of light, including the Sun, various forms of electric light, flames and materials that glow (luminescent materials). Students will distinguish between objects that emit their own light from those that require an external source of light in order to be seen

SUCCESS CRITERIA:

- identify a variety of natural light and artificial light sources
- determine which light sources give off heat and which give off no heat
- determine which light sources emit light and which reflect light
- record responses using a chart and written descriptions
- make connections to people and uses of light in the environment

MATERIALS NEEDED:

- clipboards (one per student)
- a copy of *Natural vs. Artificial Light* Worksheets 1 and 2 for each student
- a copy of *Be a Heat Detective* Worksheet 3 for each student
- a copy of *Uses of Light* Worksheet 4 for each student
- a copy of *Emission and Reflection* Worksheet 5 for each student
- pencils

PROCEDURE:

1. Engage the students in a discussion about the meaning of natural and artificial light, to ensure they have a clear understanding of what natural and artificial means. Differentiate between sources of light and reflectors of light. Give them Worksheet 1 to complete.

2. Come back together as a large group. Explain to students that some light sources give off heat and others do not. Give them Worksheets 2 and 3, and a clipboard. Be sure to review the caution warning with students to avoid unnecessary injuries. Students can work with a partner or individually to complete the tasks. One section on Worksheet 3 will need to be completed at home.

3. Give students matching activity on Worksheet 4 to complete.

4. Engage the students in a discussion about the impact of the uses of natural and artificial light to our environment, and ways we could conserve or reduce our usage. For example, street lights increase visibility and make areas safer for people to move about in the city at night, but the amount of electricity required contributes to light pollution.

5. Discuss with students the meaning of emission and reflection to ensure their understanding. Give them Worksheet 5 to complete. Come together as a large group to brainstorm ideas of what other reflectors from our everyday life that help to keep people safe.

DIFFERENTIATION:

Slower learners may benefit by working with a peer or in a small group with teacher direction in order to read through written information, to ensure understanding of content.

For enrichment, faster learners could access the internet to find photos of natural and artificial light sources. Photos could be printed out and used to create a collage. Students could then orally present collages to a small group. An extension of this could be to have the slower learners engage in an enrichment activity by identifying which objects in these collages are examples of natural light sources and which ones are artificial light sources.

Name: _____

Natural and Artificial Light

Without light from the Sun, the Earth would be cold, dark, and lifeless. Light is a valuable kind of energy for all living things. In everyday life, we get light from many different sources. Light that occurs in nature is called **natural light** and we see it in the Sun and in fires. Light that does not occur naturally is called **artificial light**. This type of light is created by people, like electric lights and electronic screens.

Let's Take A Closer Look

Circle all the sources of light that you see in the picture. Write the names of the light sources in the correct category below.

Artificial light sources: _____

Natural light sources: _____

Other things give off light too. Certain things in nature and some objects made by people can glow without being plugged in. We call material that glows **luminescent**.

Can you think of something that glows? What is it?

_____

 OTM-2173 ISBN: 978-1-4877-0201-4 © On The Mark Press

Name: _____

Light is very closely related to heat. Often things that make light give off heat. Often things that are very hot give off light.

Identify the type of light source for each picture in the chart by putting a checkmark (✓) under **Artificial Light** or **Natural Light**. Then, put a checkmark (✓) under "**gives off heat**" or "**produces no heat**".

Item	Artificial Light	Natural Light	Gives Off Heat	Produces No Heat
Glow Worm				
Candle				
Light bulb				
Campfire				
Firefly				
Flashlight				

Name: _____

Be A Heat Detective!

Take a look around your classroom or school and find objects that emit light. Test the objects to see if they give off heat as well as light.

Have an adult help you with this activity. You should always hold your hands at least 5 cm away from the light in order to feel the heat.

Name of Object	Gives off light but NO heat	Gives off light AND heat

Now try this same activity at home.

Name of Object	Gives off light but NO heat	Gives off light AND heat

 OTM-2173   ISBN: 978-1-4877-0201-4 © On The Mark Press

Name: _____

Uses of Light

Many useful items in our daily lives use light so they can function. Match the pictures of the items with their descriptions. Write the number of the sentence in the box beside the picture.

1. Mom and Dad find this very handy when they go shopping and have their hands full with bags of groceries.

2. Smile! Click, click! Light is refracted through the lens.

3. This is a source of artificial light. What we see on the screen is due to light.

4. I can easily see things far away. Light is refracted through the lens.

5. A laser light reads a code and then I have my music!

OTM-2173   ISBN: 978-1-4877-0201-4 © On The Mark Press

Name: _____

Emission and Reflection

Some objects give off their own light. They may give off artificial light, like a flashlight does, or they may give off natural light, like a firefly does. We say that they emit light. The Sun emits light.

Other objects can be seen because light bounces off them. We say that they reflect light. The Moon reflects light. We are able to see moonlight when sunlight is reflected from its surface.

Emission means _____

_____

Reflection means _____

_____

Look at the pictures. Write **emission** or **reflection** on the line beside each object to tell if it emits (gives off light) or reflects (light bounces off it).

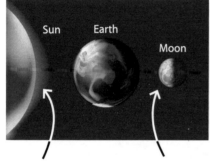 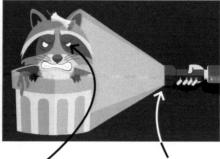 

_____ _____   _____ _____   _____ _____

_____ _____   _____ _____   _____ _____

List three reflectors from the pictures that help keep people safe.

_____ _____ _____

     OTM-2173   ISBN: 978-1-4877-0201-4 © On The Mark Press

LIGHT TRAVELS

LEARNING INTENTION:

Students will demonstrate that light travels outward from a source and continues unless blocked by an opaque material. Students will recognize that light can be bent (refracted) and that such objects as water containers, prisms or lenses can be used to show that light beams can be bent.

SUCCESS CRITERIA:

- identify paths of light and how it can be manipulated
- record observations using charts, drawings, and written descriptions
- make conclusions about the properties of light
- make connections to people, places, and the accessibility of light in the universe

MATERIALS NEEDED:

- flashlight, 3 pieces of cardboard or cardstock (a set per group of students)
- Plasticene, a jug of water, pencils
- a single hole puncher (a few would be beneficial)
- digital timers (one for each pair of students)
- clear glasses or containers (one per group of students)
- a copy of *How Does Light Travel?* Worksheets 1 and 2 for each student
- a copy of *A Change in Direction* Worksheets 3 and 4 for each student

PROCEDURE:

*This lesson can be done as one lesson or divided into shorter lessons.

1. Students will investigate how light travels. Give them Worksheets 1 and 2, and the materials to conduct the investigation. Read through the question, materials needed, and what to do sections on Worksheet 1 with students. They will conduct the investigation, record their observations, and make conclusions on how light travels.

2. Students will investigate what happens when light passes through water. Give them Worksheets 3 and 4 and the materials to do the investigation. Read through the question, materials needed, and what to do sections with them. They will conduct the investigation, record their observations, and make conclusions about the ability for light to change direction. A further discussion with students about the concept of refraction, will enhance their comprehension of the term in order for them to confidently use it as part of their scientific vocabulary.

DIFFERENTIATION:

Slower learners may benefit by working in a small group with teacher direction in order to complete the challenges on the worksheets. Chunking of information and assistance to make correct calculations will result in stronger comprehension of content.

For enrichment, faster learners could illustrate how light travels: 'Sometimes, the straight rays of sunlight are broken up by trees or clouds. Illustrate what this looks like in real life. A suggestion is to draw a forest with light showing through or big clouds letting rays of sunlight shine through.

Name: _____

How Does Light Travel?

We know that there are different sources of light. Have you ever wondered how these light sources travel? Let's investigate this idea!

Question: How do you think light travels?

Material Needed:

- a flashlight
- 3 pieces of cardboard / cardstock 15 cm x 15 cm
- Plasticine
- a single hole puncher

What To Do:

1. Make a prediction about the answer to the question, record it on Worksheet 2.

2. Cut out 3 pieces of cardboard or cardstock that measure 15 cm x 15 cm.

3. Measure to find the centre of each piece of cardboard. Mark the spot.

4. Use a single hole puncher to make a hole in the centre of two pieces.

5. Attach 2 pieces of Plasticine on the bottom edges of each piece of cardboard. Stick on enough so that the piece of cardboard will stand up by itself on a flat surface.

6. Shine the flashlight towards the pieces of cardboard. Line up the holes so that you can see the light shining through the holes.

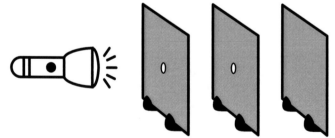

7. Now move the middle card from side to side. Observe what happens. Record your observations in the chart on Worksheet 2.

8. Make conclusions about what you observed.

OTM-2173   ISBN: 978-1-4877-0201-4 © On The Mark Press

Name: _____

Let's Predict

How do you think light travels?

_____

_____

Let's Investigate

Complete the chart with your observations.

What I Did	What I Observed
I shone the light when the holes were in a straight line.	_____ _____ _____ _____
I shone the light when the middle card was moved to one side.	_____ _____ _____ _____

Let's Conclude

Was your prediction correct?

_____

_____

What can you conclude about the way light travels?

_____

_____

_____

OTM-2173   ISBN: 978-1-4877-0201-4 © On The Mark Press

Name: _____

A Change in Direction

We have learned and demonstrated that light travels very quickly and in straight lines. But what does light do when it passes through water? Let's investigate this idea!

Question: What happens to light when it passes through water?

Materials Needed:

- a clear glass or container
- a pencil
- water

What To Do:

1. Make a prediction about the answer to the question, record it on Worksheet 7.

2. Fill the glass two-thirds full with water.

3. Place the pencil in the water. Hold it straight up.

4. Look through the glass at the pencil. Does the pencil look bent or straight?

5. Now lean the pencil against the inside of the glass. Does the pencil look bent or straight?

6. Record your observations by completing the drawing on Worksheet 7 of what you saw happen, also give a written description of your observations.

7. Make a conclusion about what you observed.

OTM-2173   ISBN: 978-1-4877-0201-4 © On The Mark Press

Name: _____

Let's Predict

What happens to light when it passes through water?

_____

Let's Observe

Complete the drawing of what you saw happen when you looked at the pencil in the water, also give a written description of your observations.

**Pencil standing
straight
up in water**

**Pencil leaning
against
the inside
of the glass**

When the pencil was standing straight up in the water I saw...

_____

When the pencil was leaning against the inside of the glass I saw...

_____

Let's Conclude

Fill in the blanks with the words and phrases in the word box below.

change directions	straight line	bend

Light travels in a _____ . But when it

passes through another medium such as water, it may

_____ or _____ .

This is called **refraction**.

THE COLOURS OF LIGHT

LEARNING INTENTION:

Students will recognize that light can be broken into colours and that different colours of light can be combined to form a new colour. Students will demonstrate the ability to use optical devices, identifying how they are used along with their general shape and structure.

SUCCESS CRITERIA:

- identify all the colours in the visible spectrum
- demonstrate how to manipulate white light to make the colours in the spectrum visible
- record observations using drawings and written descriptions
- make conclusions about the colours of light

MATERIALS NEEDED:

- a clear bowl or container, a small mirror, a piece of white Bristol board (a set per group of students), a jug of water
- sunlight
- a copy of *The Colours of Light* Worksheets 1 and 2 for each student
- a copy of *The Visible Spectrum* Worksheets 3 and 4 for each student
- a copy of *Seeing Colours* Worksheets 5 for each student
- pencils, paint, paint brushes, white paper, coloured paper, hole punchers, glue, glitter, cotton balls
- glass prisms

PROCEDURE:

1. Students will investigate what happens when light passes through a prism. Give them Worksheets 1 and 2 and the materials to do the investigation. Read through the question, materials needed, and what to do sections with them. They will conduct the investigation, record their observations, and make conclusions on how the separation of white light creates visible colours of light. (Option: this procedure can be set up as a teacher demonstration with a glass prism instead.

Students can take note of their observations on the worksheets.) A further discussion with students about the concept of a visible spectrum, will enhance their comprehension of the term in order for them to confidently use it as part of their scientific vocabulary.

2. Give students Worksheets 3 and 4. They will consolidate their learning about the visible spectrum by using the knowledge they have gained from the previous experiment to show the length of light waves in the spectrum of colour. They will connect this to the spectrum of colour that is created when sunlight rays pass through water droplets in the air.

3. Give students Worksheet 5 to complete. Read through the information about colour reflection and absorption, and discuss the concepts with students to ensure their understanding. They will need access to the internet in order to complete this activity.

DIFFERENTIATION:

Slower learners may benefit by working in a small group with teacher direction in order to complete the experiment on Worksheets 1 and 2. Seeing the colour spectrum in this experiment will give them the base knowledge to complete the subsequent activities in this section.

For enrichment, faster learners could create their own rainbow through an art activity. Ideas for art activities could be:

1) Students can make several paper dots using the hole punch in the colours of the rainbow. Then have them glue the dots onto white paper in the form of a rainbow.

2) Students can paint a picture of a rainbow, then add details such as creating clouds by gluing cotton balls to the bottom of the rainbow and adding glitter and glue to depict rain.

OTM-2173 ISBN: 978-1-4877-0201-4 © On The Mark Press

Name: _____

The Colours of Light

When light travels in a straight line, it appears to be colourless. This is called white light. But what does light look like when it passes through a prism? Let's investigate this idea!

Question: What happens to light when it passes through a prism?

Materials Needed:

- a clear glass bowl or container
- a small mirror to fit inside the glass
- water
- sunlight
- a piece of heavy white paper (Bristol board)

What To Do:

1. Make a prediction about the answer to the question, record it on Worksheet 2.

2. Fill the glass bowl two-thirds with water.

3. Place the mirror in the glass bowl or container. Place it so that it is on a slant or an angle. About one third of the mirror should be above the surface of the water.

4. Turn the glass bowl and the mirror so that they are facing the sun.

5. Hold the paper on a slant or angle in front of the mirror.

6. Move the paper until you can see a change in the light.

7. Record your observations by completing the drawing on Worksheet 2 of what you saw happen, also give a written description of your observations.

8. Make a conclusion about what you observed.

Name: _____

Let's Predict

What happens to light when it passes through a prism?

_____

_____

Let's Observe

What did you see when you moved the paper?

_____

_____

List the colours that you observed on the paper.

_____

_____

Divide the wave of light into as many sections as you observed. Colour the waves of colour in the diagram to show your observations.

Let's Conclude

Complete this statement:

When the ray of sunlight passed through this watery prism _____

_____

_____

   OTM-2173   ISBN: 978-1-4877-0201-4 © On The Mark Press

Name: _____

The Visible Spectrum

We rely on the Sun to shine each day and give us visible light. Without it we would be in darkness! The colours of the spectrum are connected to the length of the light wave. Red has the longest waves and violet has the shortest waves.

Show What You Know!

Use what you have learned from the previous experiment. Write the name of the colour of the wave length in the first box. Then colour the bar to show the wave length.

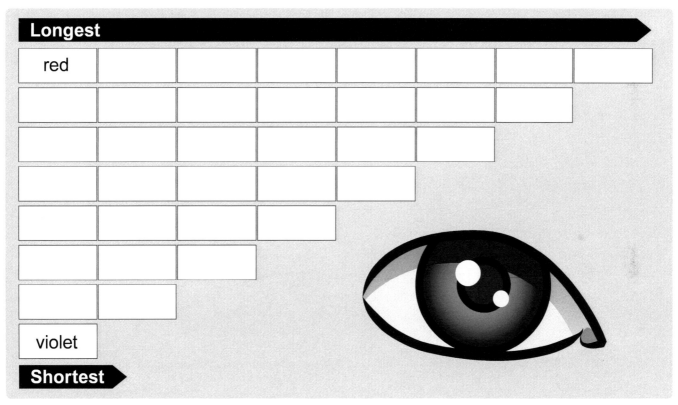

Longest

red

violet

Shortest

There are some kinds of light that we can't see because our eyes don't register light waves that are longer or shorter than the visible spectrum. Two types of light that we don't see are infrared and ultraviolet.

Use a dictionary to write a definition for these two words.

Infrared: _____

ultraviolet: _____

Name: _____

The Visible Spectrum

Think about what you know about light and the spectrum. Where have you seen this spectrum in the natural environment? In a rainbow! Read the paragraph below and use the words in the word box to fill in the missing words to learn how a rainbow is created.

spectrum	white	rainbow

A ray of sunlight contains many colours. When these colours are all

mixed together, they look _____ to our eyes. In that

white light are all the colours of a _____. We see a

rainbow when light rays pass through water droplets in the air. The

light bends and breaks up into separate colours. This is called the

_____.

Fill in the blanks to spell the colour words of the rainbow. Then add some colour to the rainbow!

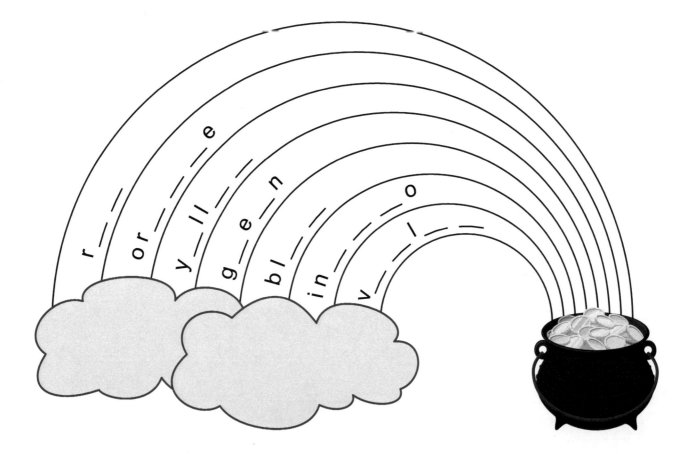

     OTM-2173   ISBN: 978-1-4877-0201-4 © On The Mark Press

Name: _____

Seeing Colours

Have you ever wondered why everyday objects are the colour they are? Most things do not make their own light. They reflect light. The colour that we see depends on the colour that is reflected into our eyes. The other colours of the spectrum are absorbed.

The colours of the spectrum are red, orange, yellow, green, blue, indigo, and violet.

What colour is reflected by:

a) a tomato _____   c) an eggplant _____

b) an avocado_____   d) a tangerine _____

e) When all the colours of the spectrum are reflected back to us, what

colour do we see? _____

Use the internet to find out what these creatures see. Put a checkmark (✓) under the correct description of how these animals see the world.

	Sees the world in many colours.	**Sees the world in black, white, and grey.**
dogs	☐	☐
cats	☐	☐
birds	☐	☐
mice	☐	☐
snakes	☐	☐

OTM-2173 ISBN: 978-1-4877-0201-4 © On The Mark Press

CASTING SHADOWS

LEARNING INTENTION:

Students will distinguish materials by examining the shadows that are cast, and classifying materials as transparent, partly transparent (translucent) or opaque. Students will demonstrate the ability to use optical devices. Students will recognize that opaque materials cast shadows as well as predict changes in the size and location of shadows resulting from the movement of a light source or from the movement of the object. Students will describe changes in the size and location of shadows made by the Sun during the day – early morning, midday and afternoon.

SUCCESS CRITERIA:

- identify and classify materials as transparent, translucent, or opaque
- determine which objects cast shadows and demonstrate how to create a shadow
- record observations using charts, drawings, and written descriptions
- make conclusions about an object's ability to cast a shadow
- make connections to people and places in the environment

MATERIALS NEEDED:

- a flashlight, coloured construction paper, white art paper, tissue paper, newspaper, page from a magazine, black garbage bag, white garbage bag, plastic wrap, plastic glass foil (a set per group of students, for the first experiment)
- a square of paper towel, a book, an overhead transparency, a flashlight (a set per group of students, for the second experiment)
- a shoe box, a solid plastic cup, a large sheet of white paper, a flashlight (a set per group of students, for the third experiment)
- a copy of *Transparent, Translucent, or Opaque?* Worksheets 1 and 2 for each student
- a copy of *What Makes a Shadow?* Worksheets 3 and 4 for each student
- a copy of *Making Shadows* Worksheets 5 and 6 for each student
- pencils, a spot light, a white sheet, a long rod, popsicle sticks, cardboard, tape, scissors

PROCEDURE:

1. Students will investigate which materials are transparent, translucent, or opaque. Give them Worksheets 1 and 2 and the materials to do the experiment. Read through the materials needed, and what to do sections with them. They will conduct the investigation, record their observations in a chart and through written descriptions, then make connections about the usefulness of transparent and opaque objects in our environment.

2. Students will investigate which materials create shadows. Give students Worksheets 3 and 4 and the materials to do the experiment. Read through the materials needed, and what to do sections with them. They will conduct the investigation, record their observations in a chart, and make conclusions about the materials that cast the best shadows.

3. Students will investigate what time of day shadows are created and when they appear longest. Give students Worksheets 5 and 6 and the materials to do the experiment. Read through the materials needed, and what to do sections with them. They will conduct the investigation, record their observations using drawings, and make connections on how the location of the Sun affects shadow casting in our environment.

DIFFERENTIATION:

Slower learners may benefit by working in a small group with teacher direction in order to complete the "Let's Conclude" and "Let's Connect It" sections for each of the experiments, to ensure students' comprehension of the classification of materials and how shadows are cast.

For enrichment, faster learners could create a play. They can cut out silhouettes of characters from cardboard. Tape a popsicle stick to the back of the silhouettes, for a handle. Suspend a white sheet as a screen between two desks, then stand between the screen and the light source with their silhouettes and perform a play.

OTM-2173 ISBN: 978-1-4877-0201-4 © On The Mark Press

Name: _____

Transparent, Translucent or Opaque?

Materials can be classified as **transparent**, **translucent**, or **opaque**.

Transparent materials transmit light and we can see right through them. **Translucent** materials transmit less light and are more difficult to see through. **Opaque** materials absorb the light around them. We cannot see through them.

Let's experiment to learn more!

Materials Needed:

- A variety of paper and plastic products:
  - coloured construction paper, white art paper, tissue paper,
  - newspaper, page from a magazine, black garbage bag,
  - white garbage bag, plastic wrap, plastic glass, foil
- a flashlight

What To Do:

1. Try to look through each paper and plastic object. Can you see through it clearly?

If you can see through it clearly, the object is _____ .

2. If you can't see through it clearly or not at all, shine the flashlight at the object. Can you see some light?

If you can see some light, the object is _____ .

If you cannot see any light, the object is _____ .

3. Record your answers in the chart on Worksheet 2.

4. Make conclusions and connections to objects in the environment.

Name: _____

Let's Observe

Record your answers in the chart. Put a checkmark (✓) under the correct category.

Object	Transparent	Translucent	Opaque
Coloured construction paper			
White art paper			
Tissue paper			
newspaper			
Page from a magazine			
Plastic wrap			
Black trash bag			
White trash bag			
Plastic glass			
Aluminum foil			

Let's Connect It!

Name three things that must be transparent in order to be useful.

_____

_____

Name three things that must be opaque in order to be useful.

_____

_____

OTM-2173   ISBN: 978-1-4877-0201-4 © On The Mark Press

Name: _____

What Makes a Shadow?

We have learned that materials can be classified as transparent, translucent, or opaque. But do all materials cast shadows?

Let's experiment to learn more!

Materials Needed:

- A variety of paper and plastic products:
  - a square of paper towel
  - a book
  - a sheet of clear plastic (overhead transparency)
- a flashlight
- a partner

What To Do:

1. Find a dark spot in your classroom with blank space on the wall.

2. Let your partner hold one of the objects about 50 cm away from the wall.

3. Shine the flashlight on the object. Did the object cast a shadow? Record your observation in the chart on Worksheet 4.

4. Repeat step 3 using the other objects.

5. Make conclusions about what you have observed.

OTM-2173 ISBN: 978-1-4877-0201-4 © On The Mark Press

Name: _____

Let's Observe

Complete the chart with your observations.

Object Tested	Transparent, Translucent or Opaque?	Did the object cast a shadow? Yes or No?
paper towel		
book		
clear plastic sheet		

Let's Conclude

Which objects failed to cast a shadow? Why do you think that happened?

_____

_____

Which objects did cast a shadow? Why do you think that happened?

_____

_____

Complete these sentences.

1. _____ objects do not create shadows

   because the light passes through them.

2. _____ objects create a dim shadow

   because some light goes through the object.

3. _____ objects create the darkest shadows

   because they absorb light.

OTM-2173   ISBN: 978-1-4877-0201-4 © On The Mark Press

Name: _____

Making Shadows

We have seen that when an object absorbs light and doesn't let it through, a shadow is cast. When we are outdoors on a sunny day, it is easy to see our own shadow.

But does our shadow look the same at all times of the day?

Let's experiment to learn more!

Materials Needed:

- A variety of paper and plastic products:
    - a solid box (shoe box)
    - a solid coloured plastic cup
    - a large sheet of white paper
- a flashlight

What To Do:

1. Place the plastic cup upside down on the sheet of white paper.

2. Shine the flashlight directly sideways at the cup. Record your observations by drawing what you see in the chart on Worksheet 6.

3. Shine the flashlight directly overhead of the cup. Record your observations by drawing what you see in the chart.

4. Repeat steps 1 to 3 using the box as your object.

5. Make conclusions about what you have observed.

Name: _____

Let's Observe

Draw your observations in the chart.

Object	**Shining the flashlight** *directly sideways*, **I saw**	**Shining the flashlight** *directly overhead*, **I saw**
plastic cup		
box		

Let's Conclude

Which angle gave no shadow? Why do you think that happened?

_____

_____

Let's Connect It!

We have learned that the size, shape, and location of a shadow depends upon the position of the light source in relation to the object. As the Earth turns, shadows change as the Sun's position in the sky changes. A short shadow at noon becomes longer throughout the day.

Draw the shadows for the cabins at each of these times.

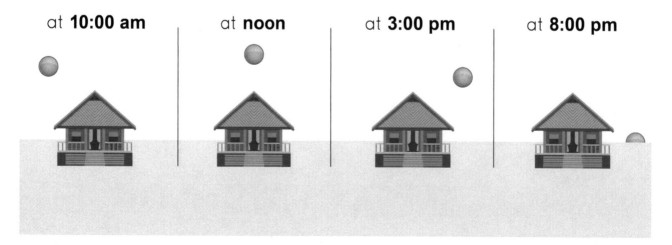

at **10:00 am**   at **noon**   at **3:00 pm**   at **8:00 pm**

OTM-2173   ISBN: 978-1-4877-0201-4 © On The Mark Press

LIGHT AND PROTECTION

LEARNING INTENTION:

Students will recognize that eyes can be damaged by bright lights and that one should not look at the Sun either directly or with binoculars or telescopes, but instead find indirect ways to study the sun.

SUCCESS CRITERIA

- identify objects used for eye protection
- practise methods intended to maintain eye protection and safety
- build a pinhole camera

MATERIALS NEEDED

- a copy of *Light and Protection* Worksheets 1, 2, and 3 for each student
- examples of tempered glass such as sunglasses and a welder's mask
- a camera, sun-spotter or other tool to look at the sun indirectly
- shoeboxes, one for each group of students, or longer boxes
- aluminum foil, masking tape, pins, scissors, flashlights to test pinhole camera
- binoculars (optional for magnifying element of a pinhole camera)
- paper, pencils

PROCEDURE:

1. Watch a video such as **https://youtu.be/ gSPDfeSgN98**, **https://youtu.be/6FB0rDsR_rc** or something similar. Review some of the material in the video with the students. Highlight the connection between light and heat. Lead students to compare what it feels like if their skin gets too hot with what might happen if their eyes see too much light.

2. Give Worksheets 1, 2, and 3 to the students. Read the material with them and monitor their work as they complete the activities.

DIFFERENTIATION:

Allow slower learners to complete the worksheets in pairs or groups and with teacher supervision. Prepare some of the material needed ahead of time for them. For example, create the pin hole in the foil or paper ahead of time.

For enrichment, faster learners could investigate how to improve their pinhole cameras by adding a magnifier, using a box of different dimensions or testing darker image surfaces within the box. Have students watch the following video (**https://youtu.be/KUAnKsW93xU**) to get ideas on how to use a binocular lens with the pinhole to magnify the overall projection of the sun.

OTM-2173 ISBN: 978-1-4877-0201-4 © On The Mark Press

Name: _____

Light and Protection

What happens if your skin gets too hot? You can burn! The same thing can happen to your eyes if they get too much light.

If you look at the sun without protection, it can cause damage to your eyes. Even if you look at the sun through regular telescopes and binoculars, you can damage your eyes. That is why it is important to use protection and keep them safe when learning about the sun.

We can still investigate the sun if we build a pinhole camera. Here's how we do it!

Materials Needed:

- a box with a separate lid
- aluminum foil
- a pin
- a clean sheet of white paper to fit in one end of the box
- a flashlight to test your device

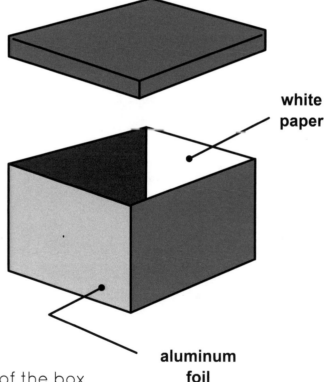

white paper

aluminum foil

What To Do:

1. Carefully cut a square out of one end of the box.

2. Wrap aluminum foil over the end of the box with the cut out square.

3. Carefully make a **small** pinhole in the foil in the middle of your cut out square.

4. Dim the lights and turn on one flashlight as a light source.

 OTM-2173 ISBN: 978-1-4877-0201-4 © On The Mark Press

Name: _____

5. Turn the box so that the pinhole side faces the light source. Then remove the top of your box and look at the opposite end of the box away from the pinhole. What do you see?

   You may have to adjust the box to centre the image.

6. Try moving the box further away from the light source. What happens?

7. Try moving the box closer to the light source. What happens?

8. Carefully try making the pinhole larger. What happens to the image?

9. Test your pinhole camera outside. You may have to keep the inside of the box very, very dark.

Options:

Turn the box upside down to keep the inside dark when you look inside.

Cut out a hole, panel or flap in the lid in order to look inside while keeping the inside dark.

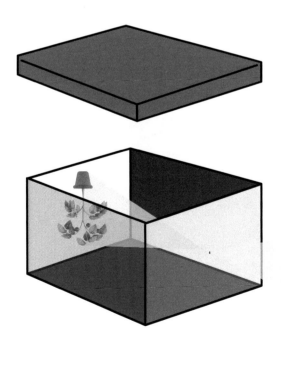

Name: _____

Let's Observe

Fill in the chart with what you found out with your pinhole camera.

What did you do?	What did you see inside your pinhole camera? Draw a picture of what you saw.	Did the image look large or small?	Was the image bright or faint?
Look in the box with a very small pinhole			
Move box farther away from light			
Move box closer to the light			
Made the pinhole larger			
Look at the Sun with your pinhole camera			

Think About It!

What could you do to make the image larger?

_____

_____

What could you do make the image brighter?

_____

_____

Is it safe to look directly at the sun through a regular telescope? **Yes   No**

Explain. _____

_____

OTM-2173   ISBN: 978-1-4877-0201-4 © On The Mark Press

TOPIC E: PLANT GROWTH AND CHANGES:
PLANT PARTS

LEARNING INTENTION:

Students will identify and describe the general purpose of plant seeds, roots, stems, leaves and flowers. Students will recognize that plants of the same kind have a common life cycle and produce new plants that are similar but not identical to the parent plants.

SUCCESS CRITERIA:

- describe the function of the root of a plant
- examine and identify plant roots as either tap or fibrous
- describe the function of the stem of a plant
- identify some edible plant stems
- describe the process of photosynthesis
- determine the different colours inside green leaves
- make observations and conclusions using written descriptions and illustrations
- identify the parts of a flower
- describe the ways the seeds of plants are distributed in nature
- identify some edible plant seeds

MATERIALS NEEDED:

- a copy of *At the Root of It!* Worksheets 1 and 2 for each student
- a copy of *Where It Stems From!* Worksheets 3 and 4 for each student
- a copy of *Plant Leaves* Worksheets 5 and 6 for each student
- a copy of *Discover the Colour!* Worksheets 7, 8, and 9 for each student
- a copy of *The Flower* Worksheet 10 for each student
- a copy of *The Seeds* Worksheets 11 and 12 for each student
- about a dozen magnifying glasses

- four different plants with root systems exposed (a set for each group of students)
- 3 glass jars, a coffee filter, 3 labels, a shallow baking pan, a wooden spoon, a collection of three types of green leaves (for each group of students)
- water, a kettle, about 4 bottles of rubbing alcohol
- access to the internet, or local library
- scissors, chart paper, markers, pencils, pencil crayons
- sheets of white art paper, crayons, differently shaped leaves (optional materials)

PROCEDURE:

*This lesson can be done as one long lesson, or divided into shorter lessons.

1. Using Worksheet 1, do a shared reading activity with the students. This will allow for reading practise and learning how to break down word parts in order to read the larger words in the text. Along with the content, discussion of certain vocabulary words would be of benefit for students to fully understand the passage.

   Vocabulary words: anchor, support, absorb, tap root, starch, tropical, nutrients, storage, fibrous root, strands

2. Divide students into small groups, and give each of them Worksheet 2, four different plants with roots exposed, and magnifying glasses. Students will name and draw each plant, and determine if each plant has a tap or fibrous root system.

3. Give students worksheet 3. Read through as a large group to ensure students' understanding. Give students Worksheet 4 to complete. They may need to visit a local library or access the internet to gather information about edible plant stems.

4. Using Worksheets 5 and 6, do a shared reading activity with the students. This will allow for

reading practise and learning how to break down word parts in order to read the larger words in the text. Along with the content, discussion of certain vocabulary words would be of benefit for students to fully understand the passage.

Some interesting vocabulary words to focus on are: trigger, broad, deciduous, evergreen, carbon dioxide, transpiration, chlorophyll, jagged, coniferous, function, photosynthesis, oxygen

5. Take students outside to collect 3 different types of green leaves (about a handful of each type). Upon returning to the classroom, divide students into small groups, and give each of them Worksheets 7, 8, and 9, and the materials to conduct the investigation.

*This investigation needs extended time to be completed, so starting early in the day is recommended. Students should conclude that the rubbing alcohol worked to break down the chlorophyll in the leaves so that their hidden colours could be revealed. This is relative to the reduction of warmth and sunlight in autumn, which results in a breakdown of the chlorophyll in nature's green leaves, to reveal their hidden colours before they die and fall from the deciduous trees.

6. Using Worksheets 10 and 11, do a shared reading activity with the students. This will allow for reading practise and learning how to break down word parts in order to read the larger words in the text. Along with the content, discussion of certain vocabulary words would be of benefit for students to fully understand the passage.

Some interesting vocabulary words to focus on are: reproductive, scent, pistil, scatter, producing, attract, fertilized, waterlogged, petals, stamen, nectar, pollen, explode, excrement, anther, distribute

7. Give students Worksheet 12 to complete. They may need to visit a local library or access the internet to gather information about edible plant seeds.

*As an activity to enhance the learning about the physical characteristics and needs of plants, show students The Magic School Bus episode called "Gets Planted". Episodes can be accessed at www.youtube.com

DIFFERENTIATION:

Slower learners may benefit by:

- working in a small group with teacher direction to complete Worksheets 7, 8, and 9.

- working with a strong peer to conduct the research on edible stems and seeds on Worksheets 4 and 12

- reducing the expectation to only finding two examples of edible plant stems and seeds on Worksheets 4 and 12

For enrichment, faster learners could do a leaf rubbing art activity. Lay a leaf under a sheet of white art paper (on a flat surface). Using the side of a crayon, lightly rub the crayon over the paper so that a pattern of the leaf can be seen. Repeat with other leaves to create a design. Red, yellow, orange, brown, and green crayons could be used as a display of the colours that are seen in autumn. An additional step would be for these students to provide a short written explanation of why these colours appear in the leaves on some deciduous trees in autumn. Then, attach it to their art work.

OTM-2173   ISBN: 978-1-4877-0201-4 © On The Mark Press

Name: _____

At the Root of It!

The roots of a plant anchor it to the ground. Roots are mostly underground and grow down into the soil. But, for some tropical trees, like the mangrove tree, some roots grow above the ground and help to support the tree.

Roots do not have leaves, but they often have root hairs which grow out into the soil to help absorb water and nutrients that a plant needs to grow. Did you know that some roots are storage areas and food for the plant? For example, the roots of beets, radishes, and carrots store food for the plant in the form of starch.

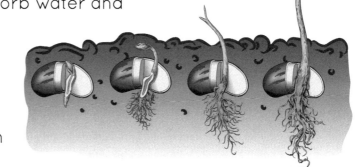

Roots are either **tap** or **fibrous**. **Tap roots** are roots that have one root that is longer than the rest. This larger root grows straight down. **Fibrous roots** have many strands of roots of similar size that spread out in all directions.

Beets, radishes, and carrots are examples of plants with tap roots. Examples of plants with fibrous roots are lettuce and grass plants.

OTM-2173   ISBN: 978-1-4877-0201-4 © On The Mark Press

Worksheet 2

Name: ______________________________

Examine the roots of four different plants. During your examination:

- name the plant
- draw the plant's root system
- determine its root system (circle tap or fibrous)

Name: ________________________

tap **fibrous**

Name: ________________________

tap **fibrous**

Name: ________________________

tap **fibrous**

Name: ________________________

tap **fibrous**

OTM-2173 ISBN: 978-1-4877-0201-4 © On The Mark Press

Name: _____

Where it Stems From!

The stem of a plant is the part that usually has buds and leaves. Stems usually grow upwards and straight, but there are some plants like the strawberry plant, which have stems that grow along the ground. The upper part of a stem of a plant can bear branches, leaves, flowers, and fruit.

Stems have four main jobs:

1. They support the plant.

2. They grow leaves.

3. They carry water and nutrients from the roots to the leaves.

4. They provide food storage for some plants.

Did you know that some stems, like tree trunks, grow a new layer every year? The new layer becomes the outside of the stem. This new layer grows underneath the bark of the tree.

The new layer is the part of the stem that brings the food and water from the roots to the leaves.

As each year passes, the stem gets thicker and thicker as new layers are added.

OTM-2173   ISBN: 978-1-4877-0201-4 © On The Mark Press

Name: _____

Did you know that some stems of plants are edible? If stems of plants provide food storage, water, and nutrients for itself, then some of those plants must be good for us to eat too!

Research It!

Find out the name of some edible plant stems. Draw and label them.

OTM-2173   ISBN: 978-1-4877-0201-4 © On The Mark Press

Name: _____

Plant Leaves

There are many different kinds of leaves that come in many different shapes and sizes. Some leaves are long and thin, while others are broad and round.

Some leaves have downy hairs on their underside. Some leaves have jagged edges, while others have smooth edges. Some leaves even have a particular smell.

The leaves on a **Venus Fly Trap** plant have stiff hairs called trigger hairs. Its trigger hairs sense insects that land on its leaves. The leaf snaps shut and traps the insect inside. It becomes food for the plant.

Leaves are often green in colour because they have **chlorophyll** in them, but these leaves have hidden colours in them too!

In autumn, the chlorophyll in these leaves breaks down and exposes the hidden colours in the leaves, such as yellow, orange, and red. The leaves then die and soon fall off. These leaves can be found on **deciduous** trees.

Coniferous trees, such as pine trees, have needle like leaves that remain green year round. Coniferous trees are also known as evergreens.

OTM-2173   ISBN: 978-1-4877-0201-4 © On The Mark Press

Name: _____

Leaves take in sunlight. They have openings that let water and air in and out. Leaves breathe in **carbon dioxide** and breathe out **oxygen**. This is very important to humans. We need the **oxygen** that the plants provide. Leaves also pass water vapor into the air. This is called **transpiration**.

The leaves on a plant have an important function for the plant. The leaves produce food for it. This food making process is called **photosynthesis**. During photosynthesis, leaves get energy from the sun, and they use their chlorophyll to create a simple sugar (**glucose**) from the air. The sugar and the water that the plant takes in make food for the plant.

     OTM-2173   ISBN: 978-1-4877-0201-4 © On The Mark Press

Name: _____

Discover the Colour!

What hidden colours are in those leaves outside? Collect some green leaves and try this!

Materials Needed:

- 3 glass jars
- a shallow baking pan
- rubbing alcohol
- a wooden spoon
- a kettle
- a pair of scissors
- a marker
- water
- a coffee filter
- 3 labels
- a collection of three types of green leaves

What To Do:

1. Tear up each type of leaf into small pieces and place each type into its own jar. Label the jars with the type of leaf that is in it.

2. Carefully pour some rubbing alcohol into each of the jars, so that it just covers the leaves that are in it.

3. Using the wooden spoon, mix and grind up the leaves in the jars.

4. Place the glass jars into a shallow baking pan. Your teacher will pour boiling water into the pan so that it covers the jars half way up.

5. After 30 minutes, remove the jars from the hot water bath. Remove the leaves from the jars. Compare the shades of chlorophyll in the different leaf types. Record your observations on Worksheet 8.

6. Cut the coffee filter into long strips. Place one strip into each glass jar so that it stands tall out of the water. Let it sit for 2 hours.

7. Pull out the strips. What do you notice? Record your observations on Worksheets 8 and 9.

8. Make conclusions about the hidden colours in green leaves. Record them on Worksheet 9.

Name: _____

Let's Observe

Part One

Describe the shades of chlorophyll in each of the jars.

Jar 1 _____ **leaves.**

_____

_____

Jar 2 _____ **leaves.**

_____

_____

Jar 3 _____ **leaves.**

_____

_____

Part Two

What did you observe after you let the coffee filter paper sit in the rubbing alcohol for 2 hours?

_____

_____

_____

_____

OTM-2173   ISBN: 978-1-4877-0201-4 © On The Mark Press

Name: _____

Illustrate the colours that you saw.

Jar 1	Jar 2	Jar 3

Let's Conclude

What did the rubbing alcohol do to the leaves?

_____

_____

_____

Explain what you have learned about the colours inside green leaves.

_____

_____

_____

_____

OTM-2173   ISBN: 978-1-4877-0201-4 © On The Mark Press

Name: _____

The Flower

Flowers are the prettiest parts of plants. They come in all shades and colours, and most of them smell very nice. Flowers also have an important job in helping a plant to survive. Let's learn more about this!

The flowers on most plants are the reproductive part. Flowers become pollinated and produce fruit, which creates the seeds to start a new plant. Some flowers go straight to seed without producing fruit, like the dandelion plant.

The petals on a plant are usually its largest part. The petals are colourful and attract animals and insects that help to pollinate the flower. The petals on a flower have a scent. The scent comes from a type of oil in the petals. The scent also attracts insects.

Inside the petals are the stamens. The stamens are long and thin. At the end of the stamen is the anther which produces the pollen.

In the very centre of the flower is the pistil. This bottle shaped part of the flower contains nectar at the top and eggs at the bottom.

Labels: stigma, pollen tube, anther, filament, stamen, petal, eggs, receptacle, sepal, pedicel, style, pistil, ovary

The eggs need to be fertilized by the pollen to produce fruit or seeds.

OTM-2173   ISBN: 978-1-4877-0201-4 © On The Mark Press

Name: _____

The Seeds

Plants reproduce themselves using seeds. Seeds come in all shapes and sizes, from the smallest, which are like the head of a pin, to the largest, which can be the size of your hand.

Seeds contain an egg that needs to be fertilized to create a new plant. Once the egg is fertilized, the plant gets the seeds ready for travel. Plants distribute many seeds, sometimes several hundred. Yet most do not take root and grow a new plant. They may get broken, or get waterlogged, or rot. Some plants even get eaten, so the seeds get destroyed.

Plants distribute their seeds in different ways. Some seeds fly with the wind, like maple keys and dandelion seeds. Some seeds have hooks and bristles that attach themselves to animals and people who walk past them. Burrs are an example of seeds that travel this way.

Other seeds travel by floating on lakes, rivers, streams, or oceans. The coconut is a fruit that often falls into the ocean. It may travel to a new shore where it breaks open and the seed within takes root. Some seeds are distributed by animals that eat the fruit of some plants. They pass the seeds as excrement.

There are plants such as impatiens that have seed pods which explode, causing their seeds to spray in all directions. Another example of an exploding seed pod is found on the rubber tree plant. Did you know that the fruit on the rubber wood tree will burst open when it is ripe, and scatter many seeds, up to about 30 metres from the tree?

These are seeds of a rubber tree plant.

Name: _____

Research It!

Did you know that the seeds of some plants are edible? Access the internet or visit your local library to help you find out the name of some edible seeds. Draw and label them.

Have you eaten any of these seeds? Which ones?

_____

_____

_____

_____

OTM-2173   ISBN: 978-1-4877-0201-4 © On The Mark Press

WHAT DO PLANTS NEED?

LEARNING INTENTION:

Students will recognize that plant requirements for growth vary from plant to plant and that other conditions such as temperature and humidity may also be important to the growth of the plant. Students will describe the importance of plants to humans and their importance to the natural environment. Students will nurture a plan through one complete life cycle, and describe the care and growth of a plant that students have nurtured.

SUCCESS CRITERIA:

- make and record predictions about the needs of a plant
- conduct experiments to investigate the needs of a plant
- make and record observations using diagrams and written descriptions
- make conclusions and connections about the needs of plants using written descriptions
- design, plan, and construct a growing environment for plants
- make and record observations of the final product by comparing and sharing similarities, differences, and benefits of the design

MATERIALS NEEDED:

*Send a note home to parents explaining that their children are going to be building an item made from cleaned, recycled materials. A suggestive list of materials to have students bring in are: egg cartons, Styrofoam trays or aluminum plates, plastic bags, boxes, plastic bottles, milk jugs, margarine tubs or other plastic containers.

- a copy of *Do Plants Need Light?* Worksheets 1, 2, and 3 for each student
- a copy of *Do Water and Air Play a Part?* Worksheets 4, 5, and 6 for each student
- a copy of *The Space Between* Worksheets 7, 8, and 9 for each student

- a copy of *A Growing Environment* Worksheets 10 and 11 for each student
- two small plants in planter containers, one labelled "light" and the other labelled "dark"
- 4 small potted plants for each group of students
- medium sized boxes (one for each group of students)
- 8 small plants, 2 planter boxes, one labelled "space" and the other labelled "no space"
- planting soil, small garden shovels, newspaper to cover desks
- access to water, a few measuring cups, a couple of watering cans
- assorted potting plants for a growing environment experiment (about two for each student)
- rulers, labels, sheets of poster paper, grid paper
- scissors, glue, string, aluminum foil, duct tape, straws, skewers
- chart paper, markers, pencils, pencil crayons

PROCEDURE:

*This lesson can be done as one long lesson, or divided into shorter lessons.

1. As a large group, conduct the experiment on Worksheet 1. Give students Worksheets 2 and 3. They will make a prediction about what may happen to a plant that is placed in direct sunlight and to a plant that is placed in a dark place. Over the next three weeks, students will make observations about the growth of both plants and record them. Then, they will make a conclusion based on their observations. (Sunlight is needed to make plants grow. Plants need the light and the heat from the sun. Sunlight also warms up the soil and rainwater.)

2. Give students Worksheets 4, 5, and 6. This experiment can be done as a large group, or in small groups. Students will make a prediction about what may happen to a plant that is placed

in a dark place with no water or air, to one that is in a dark place with air and water, to one that is in sunlight but given no water, and to one that is in sunlight, in open air, and given water. Over the next two weeks, students will make observations and conclusions about the growth of the plants and record them on Worksheets 5 and 6. Observation prompts: What colours are the leaves? Does the plant look healthy or sick? (Sunlight, water, and air are needed to make plants grow healthy.)

3. As a large group, conduct the experiment on Worksheet 7. Give students Worksheets 8 and 9. Over the next four weeks, students will make observations about the effect of adequate space on the growth of the plants. Then, they will make a conclusion based on their observations. (Space is needed by plants in order to grow healthy.)

   *Upon completion of all experiments, students should understand that plants need sunlight, warmth, water, air, and space to grow healthy.)

4. Engage students in a discussion about the importance of plants to people and other living things.

   - What do plants provide for us? (consider different points of view, for example, farmers, vegetarians, gardeners, scientists, wildlife, etc.)

   - What are the threats to plant life? (consider such things as environmental conditions, animals, and human activities)

   - What are some ways that you could reduce the harmful effects on plants and help them to survive?

5. Explain to students that they will design, plan, and construct an environment that helps plants to survive and grow healthy. Give them Worksheets 10 and 11, and the recyclable and other assorted materials to construct. Upon completion of its construction, encourage students to compare their final product with a partner's final product. They will also need to make observations about its

ability to meet plants' needs.

*As an activity to enhance the learning about the physical characteristics and needs of plants, show students The Magic School Bus episode called "Goes to Seed". Episodes can be accessed at www.youtube.com

DIFFERENTIATION:

Slower learners may benefit by:

- working as a small group with teacher support to conduct the experiment on Worksheets 4, 5, and 6 (if this experiment is not being completed as a large group activity)

- discussing their findings after each experiment within a small group with teacher support to consolidate their learning

- working with a partner to plan and design a growing environment for plants, the construction could be completed as a team or independently

For enrichment, faster learners could:

- create a "Gardening Tips" poster that defines the needs of a plant in order to grow healthy, or alternatively, create a poster that outlines the importance of caring for plants in our environment

- use grid paper to design a floor plan of a greenhouse, calculate its perimeter and area

OTM-2173   ISBN: 978-1-4877-0201-4 © On The Mark Press

Name: _____

Do Plants Need Light?

Do plants really need sunlight to grow? Let's conduct an experiment to test if sunlight affects the growth of plants!

Question: Can a plant live and grow without sunlight?

Materials Needed:

- 2 small potted plants
  (one labelled **LIGHT** and the other labelled **DARK**)
- water
- a measuring cup
- a ruler
- sunlight

What To Do:

1. Make a prediction about the answer to the question. Record it on Worksheet 2.

2. Place the **LIGHT** plant in a sunny place, and place the **DARK** plant in a dark place.

3. Every 3 or 4 days, water the plants. Be sure to give them the same amount of water.

4. Every Monday, measure the height of each plant.

5. Using Worksheet 2, record your observations of both plants every Monday, for 3 weeks.

6. Make a conclusion about what you observed.

OTM-2173 ISBN: 978-1-4877-0201-4 © On The Mark Press

Name: _____

Let's Predict!

Can a plant live and grow without sunlight?

_____

_____

Let's Investigate

This is what I saw each week.

WEEK 1	Light Plant	Dark Plant
How tall is the plant?		
Does the plant look healthy or sick?		

WEEK 2	Light Plant	Dark Plant
How tall is the plant?		
Does the plant look healthy or sick?		

WEEK 3	Light Plant	Dark Plant
How tall is the plant?		
Does the plant look healthy or sick?		

OTM-2173   ISBN: 978-1-4877-0201-4 © On The Mark Press

Name: _____

Draw and label a picture to show what happens when a plant is grown in:

Sunlight	A Dark Place

Let's Conclude

Do plants need sunlight to live and grow? Explain.

_____

_____

_____

_____

_____

_____

Name: _____

Do Water and Air Play a Part?

Question: Do water and air affect the growth of a plant?

Materials Needed:

- 4 small potted plants
  - one labelled "**DARK, AIR, and WATER**"
  - one labelled "**DARK, NO AIR, NO WATER**"
  - one labelled "**LIGHT and AIR**"
  - one labelled "**LIGHT, AIR, and WATER**"
- water
- watering can
- measuring cup
- a medium sized box
- a plastic bag

What To Do:

1. Make a prediction about the answer to the question. Record it on Worksheet 8.

2. Place the **DARK, AIR, and WATER** plant in a place without light.

3. Place the **DARK, NO AIR, NO WATER** plant in a plastic bag and place it under the box.

4. Place the **LIGHT and AIR** and the **LIGHT, AIR, and WATER** plants in a sunny place.

5. Every 3 days, water the **DARK, AIR, and WATER** plant, and water the **LIGHT, AIR, and WATER** plant. Give them the same amount of water.

6. Record your observations of all plants at the end of each week, for 2 weeks.

7. Make conclusions about what you observed.

OTM-2173   ISBN: 978-1-4877-0201-4 © On The Mark Press

Name: _____

Let's Predict

Do water and air affect the growth of a plant?

_____

_____

Let's Investigate

Describe what each plant looked like at the end of each week.

WEEK 1

DARK, AIR and WATER	DARK, NO AIR and NO WATER	LIGHT and AIR	LIGHT, AIR and WATER

WEEK 2

DARK, AIR and WATER	DARK, NO AIR and NO WATER	LIGHT and AIR	LIGHT, AIR and WATER

Name: _____

Draw a picture for each investigation and what you observed.

Drawing of the watered plant grown in the dark	Drawing of the plant grown in the dark, with no air or water
Drawing of the plant grown in sunlight, without water	Drawing of the watered plant grown in open air and sunlight

Let's Conclude

What do plants need to live and grow healthy and strong?

_____

_____

How do you know this?

_____

_____

OTM-2173   ISBN: 978-1-4877-0201-4 © On The Mark Press

Name: _____

The Space Between

We all enjoy the freedom to move and the space to stretch out. Plants are not much different than us humans. Let's conduct an investigation to see just how much plants like their own space!

Materials Needed:

- 8 small plants
- 2 planter boxes or large containers
- planting soil
- a small garden shovel
- water, watering can
- measuring cup
- a ruler

What To Do:

1. Place some soil in the planter boxes.

2. Plant 4 of the plants in the first box, very close together.

3. Plant the other 4 plants in the second box with lots of space between them.

4. Place both planter boxes in a sunny place.

5. Every 3 days, water the plants. Be sure to give them the same amount of water.

6. Record your observations of the plants in both planter boxes at the end of each week, for 4 weeks.

7. Make conclusions about what you observed.

OTM-2173   ISBN: 978-1-4877-0201-4 © On The Mark Press

Name: _____

Let's Investigate!

Describe what each plant looked like at the end of each week.

WEEK 1

Plants with no space between them	Plants with space between them

WEEK 2

Plants with no space between them	Plants with space between them

OTM-2173   ISBN: 978-1-4877-0201-4 © On The Mark Press

Name: _____

WEEK 3

Plants with no space between them	Plants with space between them

WEEK 4

Plants with no space between them	Plants with space between them

Let's Conclude

Do plants need space to grow and be healthy? Explain.

_____

_____

_____

Name: _____

A Growing Environment

Plants are important to us. Use what you know about plants to design and construct a healthy growing environment that meets all of their needs.

This is a design of my plant growing environment:

This is a list of the materials I will use to build it:

_____

_____

_____

_____

_____

_____

_____

  OTM-2173   ISBN: 978-1-4877-0201-4 © On The Mark Press

Name: _____

A Growing Environment

This is my plan for building the plant growing environment:

1. _____

_____

2. _____

_____

3. _____

_____

4. _____

_____

5. _____

_____

Now gather the materials to carry out your plan. Begin construction!

Let's Test It!

1. Plant some plants in your growing environment. Monitor them for a couple of weeks to ensure that their needs are being met.

2. Describe how your plants are doing:

_____

_____

_____

_____

SPECIAL NEEDS AND SCATTERING SEEDS

LEARNING INTENTION:

Students will identify examples of plants that have special needs. Students will describe different ways that seeds are distributed as well as recognize seed adaptations for different methods of distribution.

SUCCESS CRITERIA:

- describe how certain plants may depend on things beyond sun, water or soil-based nutrients
- explain 5 different strategies used in the dispersal of plants
- relate the special needs and seed dispersal methods of plants to seed adaptations and to the dependency of the plant on the outer world

MATERIALS NEEDED:

- a copy of *Special Plants, Special Needs* Worksheets 1, 2, and 3 for each student
- a copy of *Scattering Seeds* Worksheets 4 and 5 for each student
- pencils, pencil crayons
- access to the internet or a local library

PROCEDURE:

*This lesson can be done as one long lesson, or divided into shorter lessons.**

1. Give students Worksheets 1, 2, and 3. Read through Worksheets 1 and 2 with the students. After reading Worksheets 1 and 2 review the vocabulary and key details. Read Worksheet 3 with the students. Monitor the work of students as they complete the work. Note: some students may have to research the same plant.

Quick List of Desert Plants:	Quick List of Carnivorous Plants:
Tumbleweed	Drosera or Sundew
Saguaro Cactus	Pitcherplant
Barrel Cactus	Genlisea or Corkscrew plants
Mexican Poppies	Utricularia or Bladder Traps
Desert Sage	Large-flowered butterwort

2. Give students Worksheets 4 and 5. Read through Worksheet 4 with the students. After reading Worksheet 4, review the 5 dispersal methods, the vocabulary and key details. Read the instructions on the top of Worksheet 5. Monitor the work of students as they complete the work.

DIFFERENTIATION:

Slower learners may benefit from hands-on material and manipulatives. Bring in examples of plants and seeds to help with their drawing activities. Students may work in pairs in order to complete and share the work on time.

For enrichment, faster learners could write a short story from their desert or carnivorous plant's point of view (for example, A Day In the Life of _____.) Or, students could extend their research for Worksheet 3 and find out how their plants grow and disperse seeds.

OTM-2173   ISBN: 978-1-4877-0201-4 © On The Mark Press

Name: _____

Special Plants, Special Needs

Carnivorous Plants

Some of the strangest plants that require special needs are carnivorous plants. One of the most popular is the **Venus flytrap**. The Venus flytrap grows in the wetlands of the Southeastern coasts of the United States. Tiny hairs on the surface of the plant's leaves act as sensors. When a small insect or arachnid walks over the hairs, the pair of leaves will shut, trapping the creature.

Another example is the **Cobra lily** which grows in the west coast rainforests of North America.

These carnivorous plants tend to live in places where many of the nutrients in the soil are leached away. Like many other plants, carnivorous plants need the sun and water. But they have found a way to get the nutrients that are missing in the soil around them. They found those nutrients in small insects and spiders!

Desert Plants

The **Plains Prickly Pear** is found throughout Alberta but especially near Medicine Hat. The plant has spiny pads and a bright yellow flower. The Plains Prickly Pear grows to about 30cm tall and 80cm wide.

Like many desert plants, the Plains Prickly Pear grows in a hot and dry location. Too much water and the plant actually rots away! It prefers soil with a lot of sand mixed in so that a lot of water will drain away.

Name: _____

Special Plants, Special Needs

Research It

Look up one carnivorous plant and one desert plant.

For each one: • find the name of it

• draw a picture of it

• write a sentence explaining the special needs of the plant.

My carnivorous plant is called:

_____

Special Need:

_____

_____

_____

_____

_____

My desert plant is called:

_____

Special Need:

_____

_____

_____

_____

OTM-2173   ISBN: 978-1-4877-0201-4 © On The Mark Press

Name: _____

Scattering Seeds

If seeds fall to the ground under the parent plant, they might not get enough sun, water or nutrients from the soil. But seeds and plants can't just walk around to find more space to grow. Instead, plants have adapted to take advantage of other ways of spreading seeds. The most common methods are **wind**, **water**, **animals**, **explosion** and **fire**.

Wind

Plants like cattails, dandelions and cottonwood trees grow seeds with light and feathery bristles. The wind carries the seeds up and over long distances. The seeds end up getting blown around. They land in all kinds of places. These plants have to produce lots of seeds so that some will be successful in finding good places to grow.

Water

Mangrove trees, coconut trees and water lilies all grow beside or right in water. If a seed falls in the water, the tide or the current can carry the seed away to grow somewhere else.

Animals

Plants that grow fruit or vegetables often rely on animals. A bird, for example, may eat part of a green pepper or an orange but in the process swallow the seeds as well. When the bird flies away, it carries the seeds with them. The bird disperses the seeds in their droppings, giving the seeds all they need to start growing.

Name: _____

Some seeds have hooks or barbs. When an animal brushes past the plant, the hook can get caught in an animal's coat and take a ride. Have you ever had a bramble caught on your clothing? You've been a part of seed dispersal!

Explosion

For plants, most actions take a long time. The same thing happens for explosions with plants. Peas, radishes and flax have seedpods that slowly dry out. When the pods are dry enough, they break open to scatter the seeds.

Fire

Some pine trees produce cones that only open when they get a lot of heat. After a forest fire, the seeds from the cone start to grow. The pine cone protected the seed until the time was right to grow into a tree.

Pine cone is closed, protecting the seeds.

Pine cone is open, releasing the seeds.

OTM-2173 ISBN: 978-1-4877-0201-4 © On The Mark Press

Name: _____

Scattering Seeds

Think About It!

For each method of **seed dispersal**, Give one example of a plant that uses it. Draw a picture of the plant or the seed.

Method	Plant	Picture
Wind		
Water		
Animal		
Explosion		
Fire		

Think of a plant in your backyard or near the school. In one or two sentences, explain what its seeds look like and how it spreads its seeds.

_____

_____

_____

TOPIC A: WASTE AND OUR WORLD
ANSWER KEY

Many of the lessons in this book are experiments and hands-on activities. Students work individually, with a partner or in groups. Most student answers are based on student/child observations, opinions and conclusions. *If student results seem unclear, teachers may wish to review or reteach the concepts involved.* Definite answers are provided where applicable.

Page 10 What is Waste?

Waste is

1. leftover or thrown away materials as worthless; garbage, trash
2. used to say that something has been used up or spent in a careless way; to destroy or ruin

Think, Pair, Share

Possible Answers

1. Food: in our homes, restaurants, crops
2. Clothing: buying too much; not passing on to others
3. Forests: clearing fielsds, burning brush
4. Paper, writing supplies: schools, offices
5. Furniture: replacing it just to have new items
6. Throwing things out instead of reusing or recycling

Page 11 Human Activity vs Nature

Possible Answers

Waste Caused By Human Activity	Waste Caused By Plants and Animals
1. food waste	1. decay, bacteria that cause foods to spoil
2. clothing waste	2. green waste: plant decay that is composted
3. paper products not recycled	3. animal manure
4. computers, phones not recycled	4. tree decay caused by animals making homes
5. non-recyclable packaging	5. animals eating/stripping leaves from trees
6. excessive use of plastic	6. tree decay caused by birds digging for insects for food

Page 12 What to Do?

Student Observations and Conclusions/ Answers will vary.

Page 13 A Tree in a Forest

Possible answers

Some ways nature managed its own waste in this story are

1. drought
2. forest fire
3. ice storm
4. lightning strike
5. acid rain
6. natural decay
7. bacteria, fungi
8. high winds
9. birds, insects looking for food

Pages 14-15 The Earthworms Clean Up!

Let's Observe

Student Observations and Conclusions/ Answers will vary.

Discuss and confirm ideas.

The earthworms caused the changes in the jar by:

1. eating the foods – fruit and vegetable scraps
2. producing castings as waste
3. making tunnels in the soil

Managing Waste

Page 21 Categorize It!

Biodegradable (green box): fruit and vegetable scraps, discarded paper, wooden boards, leaves

Hazardous (orange Box): old paint cans, propane tank, medicine/pills, batteries

Recyclable (blue box): plastic containers, batteries, discarded paper, wooden boards

Page 22 Waste Disposal

Let's Investigate

OTM-2173   ISBN: 978-1-4877-0201-4 © On The Mark Press

Possible choices for method of garbage disposal: recycling, burning, landfills, composting

Student Observations and Conclusions/ Answers will vary.

Pages 23-24 Waste Management in Your Community

Let's Inquire!

Student Observations and Conclusions/ Answers will vary.

Reduce, Reuse, Recycle!

Page 27 Reducing and Reusing

Think, Pair, Share

Possible Answers

Reduce	Reuse
Food waste – buy only what you need and can use quickly; grow your own food	Compost plant food waste for soil for garden
Water: use less for household and personal uses – eg. shower instead of bath	Use water from rain barrels to water plants, garden
Less buying of foods wrapped in plastic	Newspapers, magazines
Use reusable water bottle	Glass jars, containers, cans
Bring your own bag to stores when shopping	Reuse plastic bags
Stop using plastic straws, spoons, forks, knives	Donate extra furniture to thrift stores
Repair items of clothing before buying news ones	Donate extra clothing to thrift stores

Pages 28 Recycling

Paper	Metal	Plastic	Glass
newspapers	crushed pop can	spray bottle	jam jar
gift box	food cans	pudding container	
juice box		water bottle	
		(milk) jug	

Pages 29 – 30 Let's Go Litterless!

Posters: Answers will vary.

Other ways you could advertise your litterless lunch ideas: do an in-class presentation, social media, create a contest between classes of a "least litter lunches"

Let's Predict!

Possible answers: less food waste, less packaging thrown out; people realizing how much waste we create

Let's Observe! / Let's Conclude!

Student Observations and Conclusions/ Answers will vary.

Pages 31- 33 Classroom Composting

Let's Observe!

Possible Answers

Page 32 As organic matter "cooks", the materials may get warm; steam may be seen when compost is stirred; as materials decay, the pile gets smaller

Page 33

Humus might be used to add to gardens, for potted plants

Reasons why people should compost:

Possible answers: 1) reduces waste 2) less waste going into landfills 3) reduces the amount of greenhouse gases 4) puts nutrients back into the soil 5) recycles food and garden waste 6) reduces the use of commercial fertilizers

Challenge Question: How composting happens in nature: Microorganisms from the soil eat the plant waste materials. As the plants decay, they become rich soil. A composter needs air (oxygen) so materials should be stirred

Pages 34 – 35 The Life of a Battery

1) Materials: Collect copper, aluminum, acids or electrolytes, plastic 2) Production: Use materials and energy to produce a battery 3) Final Product: Batteries are packaged in recyclable materials and put into stores 4) Using the Battery: Many things use batteries: electronic devices, smoke alarms, flashlights, toys 5) End of Use: Dead batteries should be recycled not thrown out in trash to go to landfills

Packaging and Waste

Pages 38 – 39 Pre-Consumer and Post-Consumer Waste

Student Observations and Conclusions/ Answers will vary.

Pages 40-42 What Goes Into Packaging?

Student Observations and Conclusions/ Answers will vary.

Is Packaging necessary? Possible answer: Yes because 1) we need to protect items from being damaged or broken 2) we need to prevent the spread of germs on items like food

Pages 43 – 44 Designing Your Packaging

Student Observations and Conclusions/ Answers will vary.

OTM-2173   ISBN: 978-1-4877-0201-4 © On The Mark Press

TOPIC B: WHEELS AND LEVERS
ANSWER KEY

Many of the lessons in this book are experiments and hands-on activities. Students work individually, with a partner or in groups. Most student answers are based on student/child observations, opinions and conclusions. *If student results seem unclear, teachers may wish to review or reteach the concepts involved.* Definite answers are provided where applicable.

Pages 47- 48 Pulleys All Around Us

Exercise machine: to raise and lower exercise weights

Old fashioned well: to raise and lower water pail

Crane: to raise, move and lower storage container

Gondola/Ski Lift: to transport the carrier car up and down a hill/mountain

Food belt: moves the items closer to the cashier

Elevator Belt: one belt moves the items to the elevator belt which moves it up to a higher level

Airplane Conveyor Belt: moves baggage onto and off the airplane

Belt" moves sand upward and onto the pile

Page 49 -50 Pulleys Everywhere

Student Observations and Conclusions/ Answers will vary.

Page 51 Comparing Wheels and Rollers

Toy truck: wheel

Large stone: roller

Exercise cylinder: roller

Large machinery: wheel

Pulley Systems

Pages 53-54 The Fixed Pulley: Wheel and Fixed Axle

Student Observations and Conclusions/ Answers will vary.

Yes, a pulley does make lifting easier.

Pages 55-56 Block and Tackle System: Fixed and Movable

Student Observations and Conclusions/ Answers will vary.

Gearing Up

Pages 59-61 Gears

Student Observations and Conclusions/ Answers will vary.

Gears in Motion

Page 66 Exploring the Bevel Gear

Student Observations and Conclusions/ Answers will vary.

Page 67 Exploring the Worm Gear Drive

Student Observations and Conclusions/ Answers will vary.

Pages 68-69 How Bicycles Work

Student Observations and Conclusions/ Answers will vary.

Let's Conclude

Student Observations and Conclusions/ Answers will vary.

You will need less force to turn the pedal when putting the bicycle into low gear.

You will need more force to turn the pedal when putting the bicycle into high gear.

When travelling up a steep hill, it is best to use a low gear because the resistance is less making the pedalling easier.

When travelling fast on flat ground, it is best to use middle gear because it can overcome the resistance easily.

Wheels in Motion

Page 73 Wheels in Motion

Check colouring for accuracy of answers.

Illustrations: Answers will vary.

Pages 74-75 Let's Get Rolling!

Yes, a wheel or a roller makes the moving of objects easier.

Student Observations and Conclusions/ Answers will vary.

Pages 76-77 Constructing a Car

Student Observations and Conclusions/ Answers will vary.

Pages 78-79 The Turbo Challenge!

Student Observations and Conclusions/ Answers will vary.

Levers

Pages 81-83 Levers

Confirm accuracy of labels in student diagram.

Levers Investigation Sheet

Diagram One: Effort Meter 3

Diagram Two: Effort Meter 5

Diagram Three: Effort Meter 1

The effort needed to lift the textbooks is less when the fulcrum is closer to the load.

Pages 85-86 Lever Leverage

Part One: Effort Meter 5 Check labels for accuracy.

Part Two: Effort Meter 3 Check labels for accuracy.

Part Three: Effort Meter 1 Check labels for accuracy.

The effort needed to lift the coin is considerably less when it is applied closer to the load.

Page 87 Lever Leverage! Apply the Knowledge!

Confirm accuracy of student answers. Suggestion: Discuss answers together.

OTM-2173   ISBN: 978-1-4877-0201-4 © On The Mark Press

TOPIC C: BUILDING DEVICES AND VEHICLES THAT MOVE
ANSWER KEY

Many of the lessons in this book are experiments and hands-on activities. Students work individually, with a partner or in groups. Most student answers are based on student/child observations, opinions and conclusions. *If student results seem unclear, teachers may wish to review or reteach the concepts involved.* Definite answers are provided where applicable.

Powered Up Vehicles

Pages 90-91 Balloon Power!

Student Observations and Conclusions/ Answers will vary.

Pages 92-93 Rubber Band Power

Student Observations and Conclusions/ Answers will vary.

Pages 94-95 Three Challenges: Speed, Distance and Creativity

Student Observations and Conclusions/ Answers will vary.

The Catapult (Control! Control! You Must Learn Control!)

Pages 99-100 making A Catapult

Student Observations and Conclusions/ Answers will vary.

Pages 101-102 Improving a Catapult

Student Observations and Conclusions/ Answers will vary.

TOPIC D: LIGHT AND SHADOWS
ANSWER KEY

Many of the lessons in this book are experiments and hands-on activities. Students work individually, with a partner or in groups. Most student answers are based on student/child observations, opinions and conclusions. *If student results seem unclear, teachers may wish to review or reteach the concepts involved.* Definite answers are provided where applicable.

What is Light?

Page 104-105 Natural and Artificial Light

Artificial light sources: desk lamp, camera flash, computer screen

Natural light sources: sunlight through window

Some things that glow are: firefly, electric eels, glow worm, coals in a fire

Item	Artificial Light	Natural Light	Gives Off Heat	Produces No Heat
Glow Worm		x		x
Candle		x	x	
Light bulb	x		x	
Campfire		x	x	
Firefly		x		x
Flashlight	x		sometimes	

Page 106 Be a Heat Detective!

Student Observations and Conclusions/ Answers will vary.

Television: #3 This is a source of artificial light. What we see on the screen is due to light.

Binoculars: #4 I can easily see things far away. Light is refracted through the lens.

Disk Player: #5 A laser light reads a code and then I have my music.

Electronic Door Opener: #1 Mom and Dad find this vrey handy when they go shopping and have their hands full of groceries.

Camera: #2 Smile! Click, click! Light is refracted through the lens.

Page 108 Emission and Reflection

Emission means giving off light from a source.

Reflection means bouncing light off an object

Sun: emission Moon: reflection

Animal eyes: reflection Flashlight: emission

Safety vest: reflection Spotlights on machine: emission

Lighthouse Light: emission Light on water: reflection

Headlights: emission Painted lines on road: reflection

Bike wheel reflector: reflection Light on bike: emission

Light Travels

Pages 110-111

I think light travels fast, in a straight line, through some materials

Student Observations and Conclusions/ Answers will vary.

I conclude that light travels in a straight line.

Pages 112-113 A Change in Direction

Let's Predict Answers will vary.

Let's Observe

When the pencil was standing straight up in the water, I saw the whole pencil in a straight line.

When the pencil was leaning against the side of the glass, I saw that the pencil seemed to be bent. It was not in a straight line.

Let's Conclude

Light travels in a straight line. But when it passes through another medium such as water, it may bend or change directions. This is called refraction.

Colours of Light

Pages 115-116 The Colours of Light

Let's Predict

When the light moves through the prism, it might be broken up into different colours.

Let's Observe

As I moved the paper, I saw the light being broken up into different colours.

The colours I saw are: red, orange, yellow, green, blue, indigo, violet.

Let's Conclude

OTM-2173 ISBN: 978-1-4877-0201-4 © On The Mark Press

When the ray of sunlight passed through this watery prism, it was broken into its seven colours.

Pages 117-118 The Visible Spectrum

Longest to Shortest: red, orange, yellow, green, blue, indigo, violet

Infrared: Light waves that are even longer than the deepest red.

Ultraviolet: Light waves that go beyond violet on the visible spectrum.

A ray of sunlight contains many colours. When these colours are all mixed together, they look white to our eyes. In that white light are all the colours of the rainbow.

We see a rainbow when light rays pass through water droplets in the air. The light bends and breaks up into separate colours. This is called the spectrum.

Colour the rainbow: red, orange, yellow, green, blue, indigo, violet

Page 119 Seeing Colours

Colour reflected by :

a) tomato: red b) avocado: green c) an eggplant: purple d) a tangerine: orange e) When all the colours of the spectrum are reflected back to us, we see the colour white.

Animal	Sees the world in many colours	Sees the world in black, white and grey
dogs	yellow, blue, grey	
cats	blue, grey, green	
birds	blue, green, red ultraviolet	
mice		x
snakes	blue, green	

Casting Shadows

Page 121-122 Transparent, Translucent or Opaque

What To Do

1. If you can see through it clearly, the object is transparent.
2. If you can see some light, the object is translucent.

If you cannot see any light, the object is opaque.

Let's Observe

Object	Transparent	Translucent	Opaque
Coloured construction paper			x
White art paper		x	
Tissue paper		x	
Newspaper			x
Page from a magazine			x
Plastic wrap	x		
Black trash bag			x
White trash bag		x	
Plastic glass	If clear		If coloured
Aluminum foil			x

Let's Connect It!

Possible answers

Three things that must be transparent in order to be useful: headlights on a vehicle, magnifying glass, windows

Three things that must be opaque in order to be useful: blackout drapes, aluminum foil, cardboard

Pages 123-124 What Makes a Shadow?

Let's Observe

Object Tested	Transparent, Translucent or Opaque?	Did the object cast a shadow? Yes or No?
paper towel	translucent	faded, dim shadow
book	opaque	no
clear plastic sheet	transparent	yes

Let's Conclude

The book did not cast a shadow because it is opaque and no light travelled through it.

The paper towel cast a dim shadow because it is translucent and some light passed through.

The clear plastic sheet cast a good shadow because it is transparent and much light passed through.

1. Transparent objects do not create shadows because the light passes through them.

2. <u>Translucent</u> objects create a dim shadow because some light passes through the object.

3. <u>Opaque</u> objects create the darkest shadows because they absorb light.

Pages 125-126 Making Shadows

Let's Observe

Student Observations and Conclusions/ Answers will vary.

Let's Conclude

Shining the flashlight directly overhead is the angle that gives no shadow. I think this happened because the object absorbed the overhead light.

Let's Connect It!

1) at 10:00am – shadow is on the right hand side of the house.

2) at noon – there is no shadow

3) at 3:00 pm – shadow is on the left hand side of the house

4) at 8:00 pm – little or no shadow

Pages 127-130 Light and Protection

Let's Observe

Student Observations and Conclusions/ Answers will vary.

Think About It!

To make the image larger, move the camera closer to the object.

To make the image brighter, make the pinhole larger.

No, it is never safe to look directly at the sun through a regular telescope. You should use a telescope with a solar filter.

OTM-2173   ISBN: 978-1-4877-0201-4 © On The Mark Press

TOPIC E: PLANT GROWTH AND CHANGES
ANSWER KEY

Many of the lessons in this book are experiments and hands-on activities. Students work individually, with a partner or in groups. Most student answers are based on student/child observations, opinions and conclusions. *If student results seem unclear, teachers may wish to review or reteach the concepts involved.* Definite answers are provided where applicable.

Pages 133-134 At the Root of It!

Student Observations and Conclusions/ Answers will vary.

Pages Where it Stems From!

Research It!

Student Observations and Conclusions/ Answers will vary.

Possible choices of edible plant stems: celery, asparagus, rhubarb, leeks, broccoli

Pages 139- 141 Discover the Colour!

Let's Observe

Student Observations and Conclusions/ Answers will vary.

Let's Conclude

Rubbing alcohol removes the colour and the chlorophyll from the leaves.

I learned that heat and rubbing alcohol can be used to remove colour and chlorophyll from green leaves.

Pages 143-144 The Seeds

Research It!

Possible choices for edible seeds: peas, beans, corn, nuts, lima beans

Answers will vary.

Pages 147-149 Do Plants Need Light?

Let's Predict!

Answers will vary.

Let's Investigate

Student Observations and Conclusions/ Answers will vary.

Let's Conclude

Yes, plants need sunlight to live and grow. The plants that had no sunlight didn't grow, lost their colour, withered and started to die.

Pages 150-152 Do Water and Air Play a Part?

Let's Predict

Answers will vary.

Let's Investigate

Student Observations and Conclusions/ Answers will vary.

Let's Conclude

Plants need sunlight, air and water in order to live and grow healthy and strong.

I know this because the plants that didn't have light, air and water didn't grow and they started to die.

Pages 153-155 The Space Between

Let's Investigate!

Student Observations and Conclusions/ Answers will vary.

Let's Conclude

Yes plants need space to grow and be healthy. Plants crowded too closely together do not grow properly and are not as healthy or strong.

Pages 156-157 A Growing Environment

Student Observations and Conclusions/ Answers will vary.

Special Needs and Scattering Seeds

Pages 159-160 Special Plants, Special Needs

Research It

Answers will vary.

Possible choices for a carnivorous plant: pitcher plant, cobra lily, butterwort, monkey cup, yellow pitcher plant, Australian Sundew, bladderwort

Possible choices for desert plant: cactus (many types), acacias, tumbleweed, prickly pear cactus, Joshua tree, yucca, ghost plant, red pancake

Pages 161-163 Scattering Seeds

Think About It!

Answers will vary.